Physics and Chemistry in Space   Vol. 23
Planetology

Edited by M. C. E. Huber, Noordwijk, L. J. Lanzerotti,
Murray Hill, and D. Stöffler, Münster

W0043821

Y. Kamide   W. Baumjohann

Magnetosphere-Ionosphere Coupling

With 80 Figures

Springer-Verlag

Berlin Heidelberg New York
London Paris Tokyo
Hong Kong Barcelona
Budapest

Professor YOHSUKE KAMIDE
Solar Terrestrial Environment Lab.
Nagoya University
Toyokawa 442, Japan

Dr. WOLFGANG BAUMJOHANN
MPI für extraterrestrische Physik
W-8046 Garching, FRG

Series Editors:

Professor Dr. M. C. E. HUBER
European Space Research and Technology Centre
Keplerlaan 1, NL-2200 AG Noordwijk, The Netherlands

Dr. L. J. LANZEROTTI
AT & T Bell Laboratories, 600 Mountain Avenue
Murray Hill, NJ 07974-2070, USA

Professor D. STÖFFLER
Institut für Planetologie, Universität Münster
Wilhelm-Klemm-Str. 10, W-4400 Münster, FRG

ISBN 978-3-642-50064-0   ISBN 978-3-642-50062-6 (eBook)
DOI 10.1007/978-3-642-50062-6

Library of Congress Cataloging-in-Publication Data. Kamide, Y. Magnetosphere-ionosphere coupling / Y. Kamide, W. Baumjohann. p. cm. – (Physics and chemistry in space ; v. 23) Includes bibliographical references and index.

DM 178.00 1. Magnetosphere. 2. Ionosphere. 3. Space plasmas. 4. Electrodynamics. I. Baumjohann, W. (Wolfgang) II. Title. III. Series. QC809.M35K36 1993 551.5'6–dc20 92-36356

© Springer-Verlag Berlin Heidelberg 1993
Softcover reprint of the hardcover 1st edition 1993

Typesetting: Thomson Press India Ltd., New Delhi, India
32/3145-5 4 3 2 1 0 – Printed on acid-free paper

We dedicate this monograph to

Christoph K. Goertz

one of the most innovative scientists and leaders in space physics,
a respected teacher and our best friend.
He made a real difference.

Preface

Over the last two decades, a growing number of new, direct measurements have emphasized the importance of the close electrical coupling between the earth's magnetosphere and the ionosphere. The coupling in this region implies active interactions between hot but low-density plasmas in the magnetosphere and cold but dense plasmas in the ionosphere. It is the region where acceleration of charged particles is taking place efficiently. Many phenomena observed in the polar ionosphere and many electrodynamic processes occurring in the magnetosphere cannot be physically understood unless one examines the coupled magnetosphere-ionosphere system in its entirety.

Spatial and temporal variations in high-latitude electromagnetic phenomena, such as electric fields and currents, particle precipitation, and auroras, occur as manifestations of the dynamic processes in the system and have proven to be extremely complex. Now the challenge is to comprehend and synthesize the vast amount of complicated observations made at different locations and times in our earth's environment. Recently, a significant advance has also been made through extensive computer simulation and numerical modeling in an attempt to demonstrate how the basic assumptions can reproduce the important characteristics of observations.

It is quite timely to evaluate the present status of our understanding of magnetosphere-ionosphere coupling processes by examining critically what we have learned and what we need to study in the near future. This book, however, is not an extended review article and does not pretend to be complete in any sense. Nor is it our main aim to summarize generally accepted views and treatments. This exciting field of research is expanding too rapidly for such an attempt.

It must be noted that the topics covered in this book are limited to large-scale processes in the magnetosphere-ionosphere system which are of a non-linear and time-variant nature, and within which many small-scale processes are also taking place. The intention here is to present the basic theoretical scheme of the global electrical coupling between the magnetosphere and the ionosphere, and to discuss a selection of important dynamic phenomena essential to the system. We emphasize the importance of clarifying the conditions under which specific phenomena take place and are observable by our instrumentation. It is our sincere hope that this book will further comprehension of the magnetosphere and the ionosphere. A good working knowledge of electromagnetism is required, and the reader is assumed to be familiar with phenomenology in either ionospheric or magnetospheric physics.

We thank our colleagues, who are too numerous to mention individually, for their useful discussions, conversations, and words of encouragement. We would like to acknowledge Lou Lanzerotti, who persuaded us to contribute to the series, *Physics and Chemistry in Space*. Those who were especially helpful by providing critical comments on an earlier version of the final manuscript include J. R. Kan, A. Nishida, A. D. Richmond, and R. A. Wolf, without whose efforts the book would not have taken its present form/contents. Special thanks are due to the High Altitude Observatory of the National Center for Atmospheric Research, under whose auspices and hospitality portions of the final manuscript were prepared.

January 1993

Y. KAMIDE
W. BAUMJOHANN

Contents

1 Implications of Magnetosphere–Ionosphere Coupling

The Earth's magnetosphere is a vast laboratory for plasma physics. It is bounded by the magnetopause and solar wind on the outer side and by the ionosphere on the inner side. These boundaries are not stationary topographical borders but dynamic regions, where important plasma parameters are different from those inside the magnetosphere and where unique physical processes operate.

Since the magnetopause, the magnetosphere and the ionosphere are threaded by the same magnetic field lines, momentum and energy are exchanged among these different regions. Therefore, dynamic processes occurring in each of these regions have to interact and adjust to each other, creating a new type of physics.

In order to understand the global behavior of the magnetosphere, it is necessary to view the solar wind–magnetosphere–ionosphere system as a whole. However, since it is difficult to have a complete overview of the coupling between the magnetosphere and one of its boundaries, one still tends, at the present stage, to divide the system in order to deal either with the solar wind–magnetosphere coupling processes or with the interactions between the magnetosphere and the polar ionosphere.

1.1 Solar Wind, Magnetosphere and Ionosphere

The Earth's magnetosphere is a cavity filled with hot, but dilute plasma embedded in the fast-flowing denser, but colder solar wind plasma. Due to the magnetic field originating in the Earth, the solar wind cannot directly penetrate the outer boundary of the magnetosphere, the magnetopause, but is deflected around it after having been slowed down to subsonic velocities at the Earth's bow shock (e.g., Russell 1987). As shown in Fig. 1.1, the kinetic pressure of the shocked solar wind compresses the dipolar terrestrial magnetic field on the dayside and, on the nightside of the Earth, transforms it into a long, tail-like structure, which reaches far beyond the lunar orbit. At greater distances the magnetotail field no longer resembles a dipolar field, but rather consists of two separate regions of nearly antiparallel fields. The transition region is often called the neutral sheet.

The central part of the magnetotail, in the neighborhood of the neutral sheet, is populated by plasma with energies of several kiloelectronvolts (keV). This region, the plasma sheet, is connected with the auroral ionosphere along magnetic field lines. Closer to the Earth energetic ions, with energies of some tens of keV, bounce back and forth between the converging fields in the northern and southern ionosphere and drift westward under the influence of gradient and curvature drift. The westward drift constitutes a considerable current encircling the Earth; accordingly, this region is called the ring current.

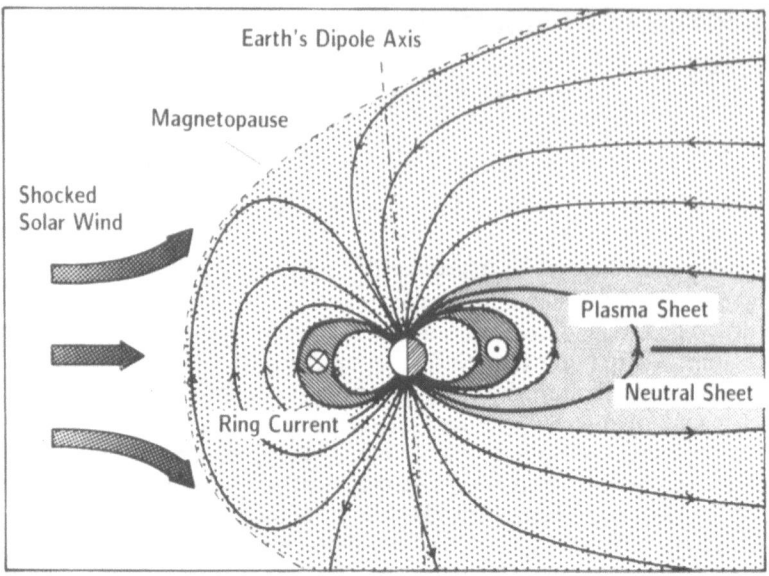

Fig. 1.1. Topography of the Earth's magnetosphere in a noon-midnight meridional cross-section

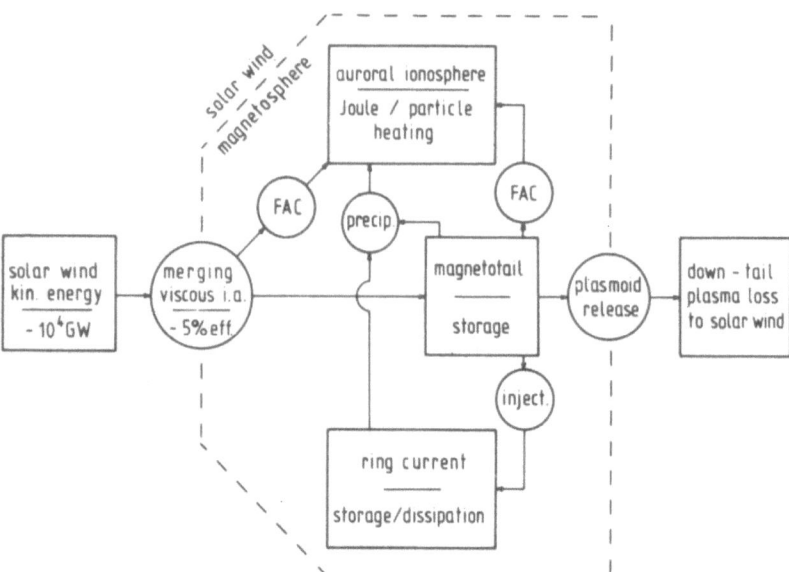

Fig. 1.2. Schematic diagram of the flow and dissipation of solar wind energy entering the Earth's magnetosphere; FAC denotes field-aligned current. (After Baumjohann 1986)

The kinetic energy flux of the solar wind, impinging on the dayside magnetopause with its cross-section of about $30 R_E$ (Earth radii), can be estimated to be about 10^4 GW for typical solar wind conditions. A fraction of this energy flux is extracted by several different mechanisms and enters the Earth's magnetosphere where it is distributed, intermediately stored, and eventually released into and dissipated in the inner and outer magnetosphere and the polar ionosphere. The entire process is shown schematically in Fig. 1.2.

Some of the energy entering the dayside magnetosphere is directly transferred to the high-latitude ionospheres via field-aligned currents where it is dissipated by the Joule heating. The remainder is transferred to the magnetotail where it is intermediately stored in the form of kinetic and thermal plasma energy as well as magnetic field energy. Part of the stored magnetotail energy is then transferred to the ring current via particle injection. The ring current energy is dissipated mainly by charge exchange with neutrals and subsequent loss to the Earth's atmosphere. Another part of the energy stored in the magnetotail is transferred to the auroral ionosphere via particle precipitation and field-aligned currents and dissipated in the form of Joule and particle heating. The remainder of the intermediately stored magnetotail energy is not dissipated inside the magnetosphere but re-enters the downstream solar wind through the down-tail release of plasmoids.

According to order-of-magnitude calculations, it is estimated that during strongly disturbed periods about 800 GW are dissipated, at about equal amounts, in the three aforementioned regions, i.e., the polar ionospheres, the ring current, and the distant magnetotail. Under more quiet conditions the dissipation rate is around 200 GW. Comparing the impinging energy flux of the order of 10^4 GW with the latter numbers for energy dissipation, the solar wind-magnetosphere coupling has an efficiency of about 5%. This 5% of the energy flux is conveyed to the magnetosphere by several different transfer processes. In the next section we will evaluate their relative importance, i.e., their contribution to the overall coupling efficiency. In Section 1.1.2 we will direct our attention to the process by which a significant amount of energy is dissipated. This is the magnetospheric substorm.

1.1.1 Entry of Energy into the Magnetosphere

The energy coupling processes at the magnetopause can be divided primarily into two different categories: (1) those processes called magnetic reconnection or field line merging, which imply interaction between the solar wind magnetic field, i.e., the interplanetary magnetic field (IMF), and the terrestrial field at the dayside magnetopause (Dungey 1961); (2) other "non-magnetic" mechanisms which one usually refers to as viscous-like interactions since they imply that tangential momentum is transferred from the magnetosheath plasma through the magnetopause via some kind of viscosity generated by micro- or macro-instabilities (Axford and Hines 1961). Figure 1.3 illustrates these two processes and the resulting plasma flow in the magnetosphere. Magnetic reconnection will drive a tailward plasma flow on open field lines across the polar caps and the

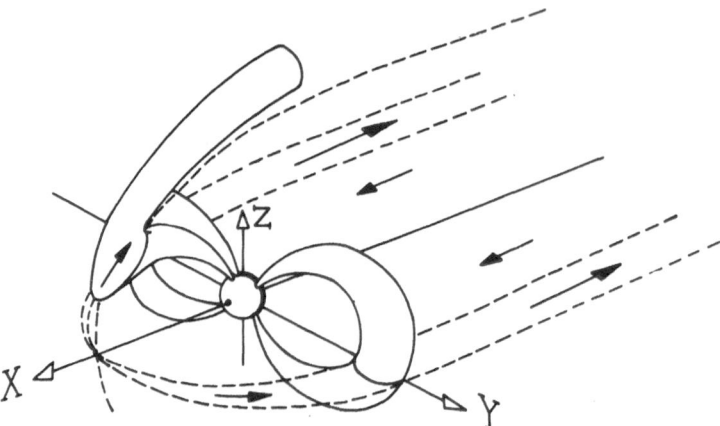

Fig. 1.3. Illustration of the two basic processes contributing to solar wind-driven convection in the magnetosphere: reconnection and viscous-like interaction. The X-axis points toward the sun, the Y-axis toward dusk. (Cowley 1982)

magnetospheric lobes, while viscous-like processes will drive a tailward plasma flow in the low-latitude boundary layers which are threaded by closed field lines (e.g., Eastman et al. 1976). For both processes the convection cycle will be completed by sunward convection in the inner magnetosphere.

Magnetic reconnection at the Earth's magnetopause (e.g., Vasyliunas 1975) operates in both a steady-state and an intermittent time-dependent fashion. Satellite missions have provided in situ evidence for quasi-steady reconnection in the vicinity of the subsolar point (e.g., Paschmann et al. 1979) and for patchy, impulsive reconnection, so-called flux transfer events (e.g., Russell and Elphic 1979). Reconnection is a process that has a strong dependence on the angle between the interplanetary magnetic field and the terrestrial field: it operates most efficiently for antiparallel field orientation and ceases for purely parallel fields. Reconnection can provide the transfer of mass and magnetic flux (or electric potential) of solar wind origin through the magnetopause, which is required to replace those particles lost by precipitation into the atmosphere and to drive an average level of magnetospheric convection (Hill 1983).

Viscous-like interaction is comprised mainly of two mechanisms: diffusion of magnetosheath particles through the magnetopause into the low-latitude boundary layer via stochastic scattering at resonant waves generated by micro-instabilities (e.g., Tsurutani and Thorne 1982) and the Kelvin-Helmholtz instability (a macro-instability) at the flanks of the low-latitude magnetopause, especially when evolving into the non-linear regime (e.g., Miura 1984). Both processes can operate independently from the external magnetic field orientation, but neither of these two processes alone is capable of fulfilling the observational requirements of mass transfer and generated electric potential. Diffusion may, in principle, become marginally competitive with magnetic coupling if it were to proceed at its maximum conceivable rate (Hill 1983), but even the upper limit for the generated potential is still below the empirically required value. In its non-linear stage, the Kelvin-Helmholtz instability can provide a potential of the order of the empirical requirement (Miura 1984), but fails to provide any mass transfer without additional diffusion or reconnection processes.

As mentioned above, one of the most pronounced and easily observable differences between reconnection and viscous-like interaction is their dependence or independence, respectively, on the direction of the external magnetic field as expressed by the sign of the B_z component of the IMF. This dependence on the interplanetary B_z component has been used over the last 20 years in numerous correlation studies between IMF changes and geomagnetic activity (see Murayama 1982; Maezawa and Murayama 1986; and the summary table in Akasofu 1981). The most reliable measures on the relative importance of reconnection and viscous-like interaction were obtained from correlations between the interplanetary magnetic field and the magnetospheric or ionospheric electric field.

Figure 1.4 shows a distinct difference in the magnetospheric convection pattern at a radial distance of $6.6R_E$ between periods when IMF $B_z < 0$ and periods of northward B_z (Baumjohann and Haerendel 1985). If viscous-like interactions would dominate magnetospheric convection, the plasma flow at synchronous orbit should be sunward directed regardless of the IMF direction. One sees, however, sunward flow for negative B_z (when conditions for reconnection are favorable) and more or less clockwise plasma circulation during periods of northward B_z. During the latter intervals, the solar wind dynamo obviously works rather inefficiently and the convection in the Earth's magnetosphere is strongly influenced by ionospheric dynamo action.

Results along the same line were obtained earlier for the correlation between the cross-polar cap potential Φ_{pc}, measured by low-altitude satellites, and the east-west component of the interplanetary electric field, i.e., the north-south

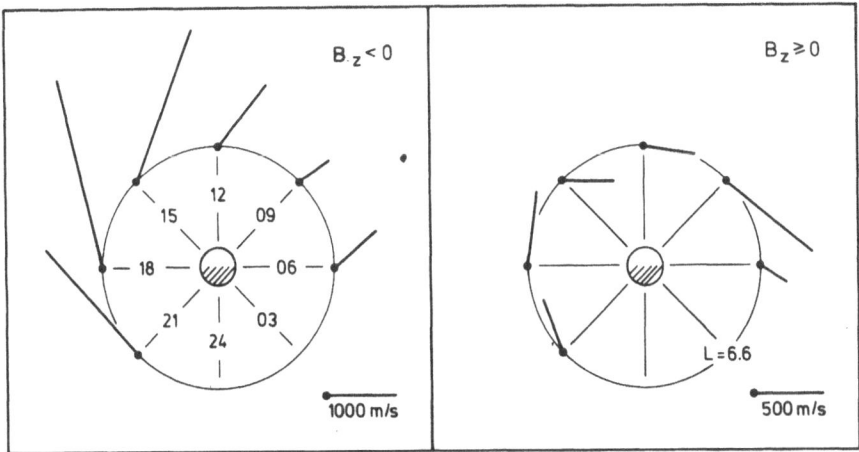

Fig. 1.4. $\mathbf{E} \times \mathbf{B}$ plasma drift vectors (in the co-rotating frame of reference) at $L = 6.6R_E$ measured by the GEOS-2 electron gun experiment for positive and negative IMF B_z components. (Baumjohann and Haerendel 1985)

component of the IMF (e.g., Reiff et al. 1981; Rich and Maynard 1989; Weimer et al. 1990). All of these studies show a significant dependence of Φ_{pc} on the IMF B_z component, indicating that the observed changes in Φ_{pc} are reconnection-associated. It is important, however, to point out that the studies also indicate the existence of substantial cross-polar voltages when the simultaneous hourly averaged B_z is positive. After taking into account the solar wind and IMF conditions during the four previous hours, Wygant et al. (1983) placed an upper limit of 16 kV on the unmodulated component of the flow assumed to be associated with viscous-like processes.

Using in situ observations made by the ISEE satellites, Mozer (1984) calculated the potential across the dusk-side, low-latitude boundary layer from measurements of the electric field component normal to the adjacent magnetopause. The resulting average potential was of the order of 2.5 kV regardless of the sign of the magnetosheath B_z component. Assuming a similar boundary layer potential at the dawn side yields an average viscous potential of about 5 kV, even less than the upper limit of Wygant et al (1983). Since typical cross-polar cap potentials have values of 50–100 kV, one has to conclude that viscous-like interaction processes cannot provide more than 10% of the total potential required to drive magnetospheric convection. On the other hand, reconnection processes seem capable of providing the empirically required voltages, in addition to displaying the aforementioned modulation with the IMF B_z component.

We will leave the discussion of the solar wind-magnetosphere-coupling at this stage. The reader interested in a more extensive overview on this topic is referred to reviews by Haerendel and Paschmann (1982), Lee (1986), Baumjohann and Paschmann (1987), and Siscoe (1988).

1.1.2 Dissipation of Energy in Substorms

The energy entering the Earth's magnetosphere through the various transfer processes described in the previous section must, of course, be dissipated somewhere in the system. While this can be done in a quasi-steady manner when the magnetopause energy transfer rate is low, the magnetosphere reacts in a non-stationary or explosive manner when the energy input rate is relatively high. The term magnetospheric substorm comprises the most basic phenomena by which the magnetosphere tries to adjust itself to enhanced solar wind input, especially following periods of southward interplanetary magnetic field.

The substorm is a global phenomenon whose manifestations can be noted in nearly all regions of the magnetosphere and the ionosphere; see, for example, Akasofu (1977), Rostoker et al. (1980), and McPherron (1991). At the ionospheric level the aurora breaks up, electric fields are distorted, and strong currents appear (e.g., Baumjohann 1983). At synchronous orbit, just outside the ring current region, the terrestrial field is first stretched out into a tail-like configuration and then suddenly switches back to dipolar (e.g., McPherron 1979), concurrent with the dispersionless injection of hot plasma into the inner magnetosphere. Further out in the tail the plasma sheet first thins and then expands again (e.g., Hones et al. 1971). The plasma is heated considerably and high-speed flows appear (e.g.,

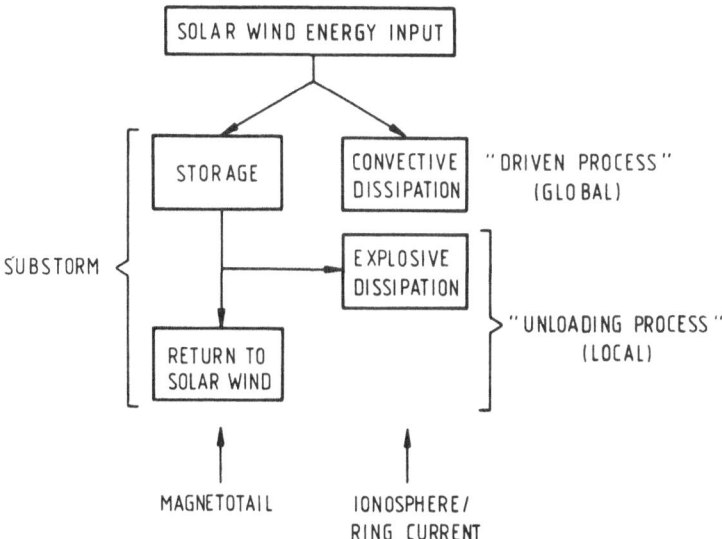

Fig. 1.5. Illustration summarizing the storage and dissipation of solar wind energy by both driven and loading-unloading processes during a magnetospheric substorm. (After Baker et al. 1984)

Baumjohann et al. 1989). In the distant tail closed magnetic structures, so-called plasmoids, are ejected down-tail (e.g., Scholer et al. 1984).

Nowadays there is widespread agreement that the magnetospheric substorm goes through two basic concurrent processes, the directly driven process and the loading-unloading process (Akasofu 1981; Baker et al. 1984; Rostoker et al. 1987b). As schematically illustrated in Fig. 1.5, some part of the enhanced solar wind energy input is directly dissipated by means of global convection, leading to Joule heating of the auroral ionosphere by the enhanced convection currents and deposition of particle energy in the auroral ionosphere as well as in the ring current. This energy dissipation is directly correlated with the solar wind energy input and constitutes a driven process, i.e., directly driven by the solar wind (e.g., Perreault and Akasofu 1978). The remainder of the enhanced energy input is, at the same time, stored intermediately in the Earth's magnetotail and then, at substorm onset, is rather explosively released via Joule and particle heating of localized regions of the auroral ionosphere, as well as by injection of particles into the ring current and by the down-tail release of plasmoids. This second substorm process, which operates concurrently with the driven process, has been named the loading-unloading process.

Figure 1.6 provides both a schematic diagram and some data on the concurrent operation of the driven and the loading-unloading processes during and after a period of enhanced solar wind input (Baker et al. 1985). Soon after the energy coupling between the solar wind and the magnetosphere is enhanced due to a southward interplanetary magnetic field, energy is stored in the magnetotail. With a time delay of about 10–20 min due to the inductance of the system, energy is directly dissipated in the auroral ionosphere in the form of Joule heat. The

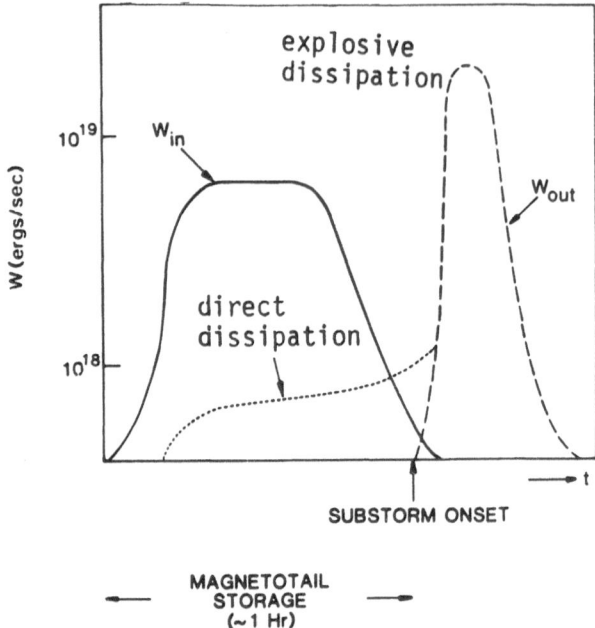

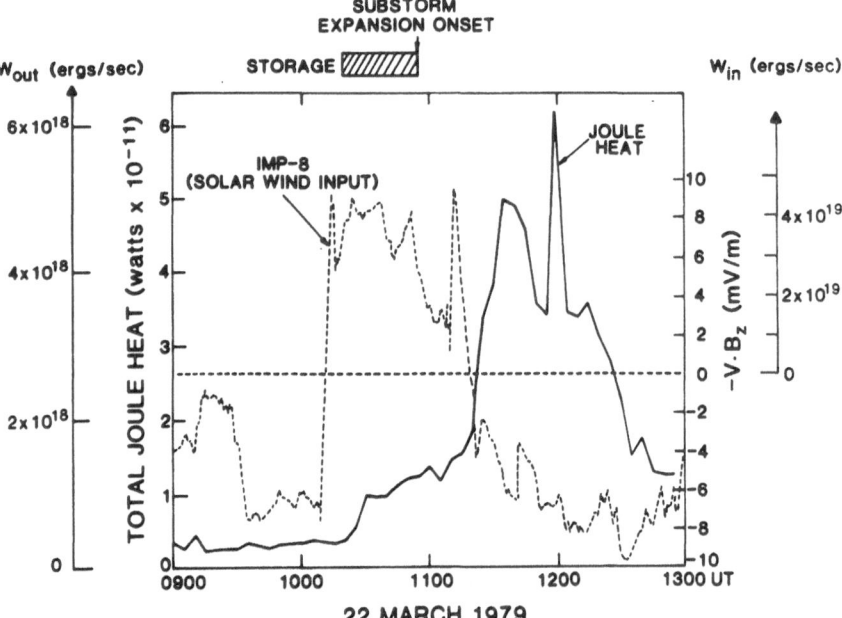

Fig. 1.6. Schematic diagram and data describing the direct dissipation (driven process) and the concurrent magnetotail storage and explosive dissipation (loading–unloading process) during a magnetospheric substorm. (After Baker et al. 1985)

stored energy is then, typically after about 30–60 min, explosively released during the expansion phase, often but not necessarily after a reduced solar wind input.

It should, however, be noted that in contrast to the impression one obtains from Fig. 1.6, direct and explosive dissipation of energy can coexist during the expansion phase. Furthermore, the question as to which of these two processes dominates, i.e., dissipates more energy during a substorm, cannot be answered easily. In the case shown in Fig. 1.6 explosive dissipation is clearly dominant, but other cases have been observed in which the dominant energy dissipation had to be attributed to the driven process (e.g., Pellinen et al. 1982).

The substorm is still a topic of much debate, thus there is as yet no unified substorm model, but there are several competing models which explain certain facets and manifestations of this complicated process. In the most comprehensive

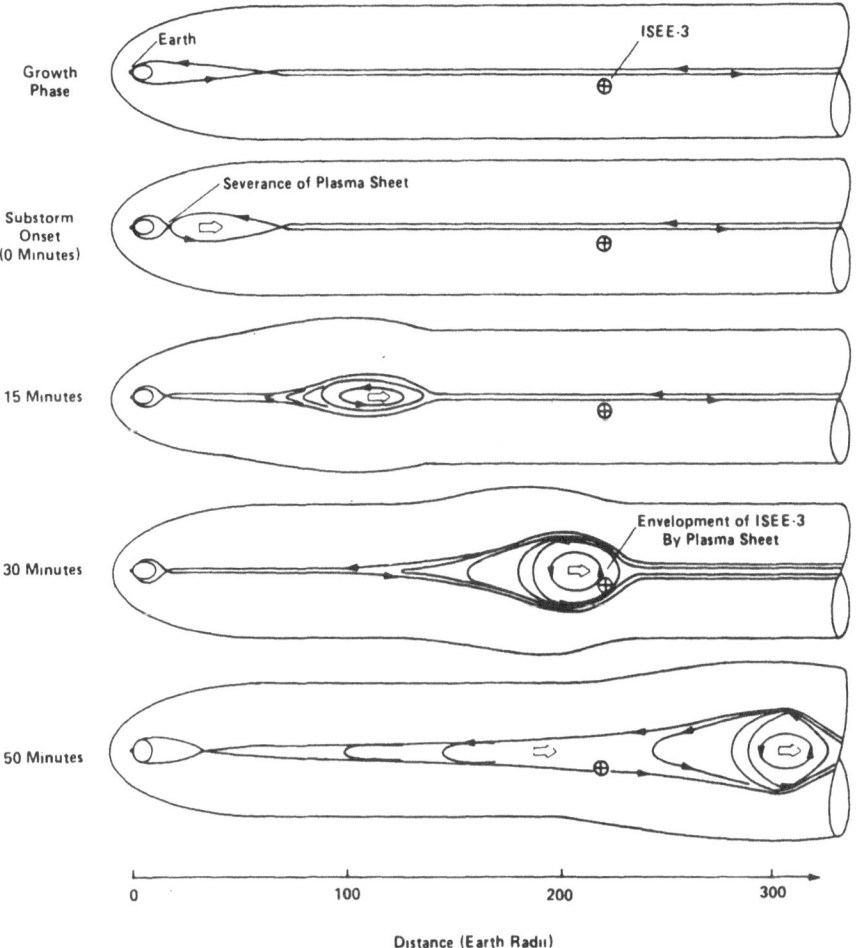

Fig. 1.7. Schematic diagram showing the tailward release of a plasmoid. (After Hones et al. 1984)

model, the near-Earth neutral line model, the unloading of tail energy is initiated by the formation of a near-Earth neutral line, where magnetic reconnection converts the stored magnetotail energy. The instability which most likely leads to the onset of near-Earth reconnection is the ion tearing mode.

Results from the ISEE-3 deep-tail satellite mission have added considerable support to this substorm picture through the unambiguous observation of tailward travelling, closed magnetic structures, the so-called plasmoids (e.g, Hones et al. 1984b; Scholer et al. 1984; Slavin et al. 1989). As shown in Fig. 1.7, these plasmoids are created at substorm onset between the newly formed near-Earth neutral line and the presumably always existing distant neutral line. Due to the slingshot effect (magnetic tension), these plasmoids are then ejected down-tail and engulf the ISEE-3 satellite at about $220R_E$ down-tail some 30 min after substorm onset.

The typical total kinetic, thermal, and magnetic energy contained in a plasmoid was estimated to be of the order of 10^{15} J (Scholer et al. 1984). This is about the same amount of energy typically dissipated in the auroral ionosphere during the course of a substorm with an average magnitude (Baumjohann 1986). Furthermore, since the average amount of energy deposited in the ring current during the course of a substorm is of the same order of magnitude as ionospheric heating (Baumjohann and Kamide 1984), one may say that the energy released in a substorm is roughly equipartitioned between the three different deposition regions, namely, the auroral ionosphere, ring current, and down-tail solar wind.

As in the previous section, we can only describe the most basic features of the substorm at this time. Detailed information on substorm-associated fields and current is given in the next chapter. A more complete picture of the morphology and physics of the magnetospheric substorm can be found in reviews by, for example, Baker et al. (1984), Baumjohann (1986, 1988), and McPherron (1991).

1.2 Basic Properties of Magnetosphere–Ionosphere Coupling

The Earth's magnetosphere is dominated by a collision-free plasma, while the ionosphere is the region where the effects of collisions of charged particles with neutral particles cannot be neglected and electrical conductivities transverse to the geomagnetic field maximize. The magnetic field connects electrically the ionosphere and the magnetosphere, causing an exchange or coupling of energy and momentum between the two regions. In a sense, the magnetosphere–ionosphere coupling is the interaction of different physical processes taking place in either of these two regions.

The strong coupling occurs since the two regions are connected by the same magnetic field lines. This interaction is important for many phenomena such as plasma convection in both the magnetosphere and the polar ionosphere, magnetospheric substorms, the formation of auroral potential structures through which the acceleration of auroral particles and the corresponding generation of auroral forms occur, the penetration of high-latitude electric fields into low latitudes, and the flow of field-aligned currents.

1.2.1 Global and Local Coupling Processes

The fundamental equation of motion of plasma particles is

$$m_j \frac{d^2 \mathbf{x}_j}{dt^2} = q_j \, (\mathbf{E} + \mathbf{v}_j \times \mathbf{B}), \tag{1.1}$$

where m_j, q_j, $\mathbf{x}_j$, and $\mathbf{v}_j$ are the mass, charge, position, and velocity of a plasma particle specified by the subscript j, and $\mathbf{E}$ and $\mathbf{B}$ are the electric and magnetic fields, respectively. If the electric and magnetic fields were fixed in space and time, one could solve (1.1) for all particles. There are, however, inherent problems in our area of space physics, particularly in magnetosphere–ionosphere coupling processes, because the positions and velocities of the plasma particles change or even create the fields in (1.1). Often plasma particles as well as neutral particles even interact with each other, which would make it necessary to include collision terms in (1.1); for the collision terms see Eqs. (3.1) and (3.2). All this makes it difficult to solve (1.1). In other words, the system is highly non-linear.

It is important to realize that (1.1) governs both microscopic (or local) and macroscopic (or global) processes in space plasmas. Although our goal is to understand energy flow and conversion in both the microscopic and macroscopic sense, these two types of processes are very different in their temporal behavior and the size of their associated phenomena. Taking typical magnetospheric conditions, the microscopic temporal-spatial scales differ more than several orders of magnitude from those of macroscopic processes (Sato 1982). For example, the ratio between the characteristic global length (say, 30 000 km) and the electron Debye length (about 500 m) reaches 6×10^4. Here, the Debye length in meters is defined as $\lambda_D = 7.43 \sqrt{nT}$, if the electron temperature T is given in eV and the electron density n is given in cm^{-3}.

The microscopic, local processes are the result of energy conversion between particles and waves, generating, in the magnetosphere–ionosphere system, local acceleration and radio wave emissions. To solve these microscopic processes, the so-called kinetic approach is required. On the other hand, macroscopic, global processes occur as a result of large-scale energy conversion. A good example of such macro-/microscopic interactions is the magnetospheric substorm described in Section 1.1.2.

This book deals primarily with the macroscopic processes, although microscopic processes will also be mentioned whenever necessary, for example, when discussing the formation of auroral arcs. It must be noted that, strictly speaking, both microscopic and macroscopic processes must be simultaneously accounted for to fully understand our electrodynamic environment, because the two types of processes, such as particle acceleration (most likely a microscopic process) and the generation of field-aligned currents (a macroscopic process), are taking place in the same region at the same time.

1.2.2 Plasma Convection

The magnetosphere and the ionosphere are coupled in so many different and complicated ways that the subject of the magnetosphere–ionosphere coupling system is notoriously difficult to subdivide practically. All magnetospheric plasma processes are in some way controlled by electrodynamics in the ionosphere and all ionospheric processes pertain in some way to the multifaceted state of the magnetosphere.

Plasma convection in the magnetosphere is one of the major aspects of macroscopic dynamic processes in the magnetosphere and is important in controlling physical processes in the magnetosphere–ionosphere system. The large-scale pattern of plasma motion and electric fields in the magnetosphere and its consequences were described theoretically by Axford and Hines (1961) and Dungey (1961). In the early 1960s, practically no direct measurements of the magnetospheric properties had been made. However, the essential features of magnetospheric convection, suggested the above authors, using the magnetohydrodynamic approximation (MHD, i.e., $E + v \times B = 0$, where E is the electric field, B is the magnetic field, and v is the bulk velocity of plasma), are still valid, indicating that the concept of magnetospheric convection can be based primarily on theoretical models using the MHD approach.

These qualitative theories were made more quantitative by Fejer (1964), Swift (1967, 1968), and others. Magnetospheric convection and the associated electric fields and currents were further developed with emphasis on the dynamics in the magnetosphere–ionosphere coupling system, which researchers have been attempting to understand with increasing sophistication (Nishida 1966; Axford 1969; Atkinson 1970; Wolf 1970, 1975; Swift 1971; Taylor and Perkins 1971; Vasyliunas 1972; Jaggi and Wolf 1973; Pudovkin 1974; Boström 1975; Gurevich et al. 1976; Rostoker and Boström 1976; Southwood 1977; Stern 1977; Sato 1978). In the last decade modeling efforts became more computer-oriented due to the limitations of analytic calculations to account for the increasing number of observations, indicating a rather complex spatial and temporal structure of the parameters governing the set of equations.

1.2.3 Theoretical Approach

The self-consistent logic governing the entire magnetosphere–ionosphere coupling system, along with the elements constituting the system, are shown in Fig. 1.8 (Vasyliunas 1970a). This block diagram shows only the core of the system, but there are a number of other observable physical parameters within the system. Because of the close coupling between the two regions, i.e., the magnetosphere and the ionosphere, we can achieve an understanding of the physical mechanisms only if we examine the coupled system in its entirety.

In general, if the magnetospheric electric field as well as the magnetic field configuration are known, the particle motion can be calculated, whereby the ion and electron populations display very different dynamic behaviors. In other words, the distribution of the plasma pressure and the particle drift current in the magnetosphere can be estimated. For practical computations, the isotropic pitch-angle approximation (Kennel 1969) is used. That is, it is assumed that the pitch angles of particles under consideration are scattered many times in the time it takes them to drift a significant distance in the magnetosphere. The magne-

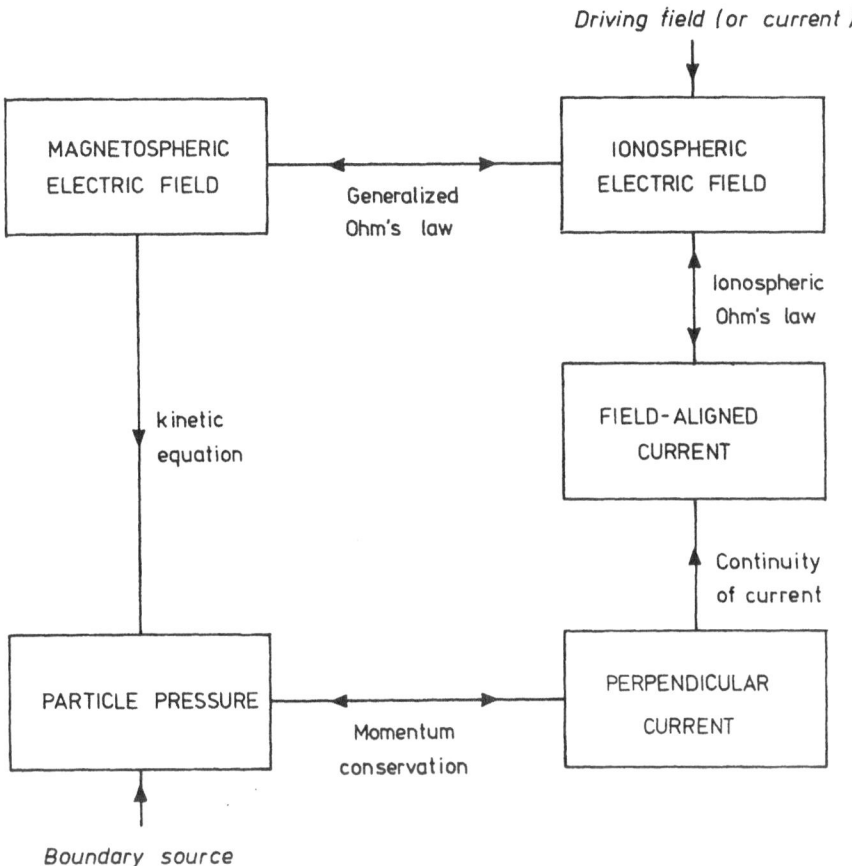

Fig. 1.8. Logic diagram for a self-consistent treatment of electrodynamic processes in the coupled magnetosphere–ionosphere system. The *boxes* indicate quantities to be determined and each *line* joining two boxes is labeled with the physical principle that governs the relation between the two quantities. (Vasyliunas 1970a)

tospheric particle population is characterized in terms of energetic ions and electrons as well as cold plasma.

The current density, **J**, in the magnetosphere is related to the plasma pressure, p, and the plasma velocity, **v**, as

$$\rho \frac{d\mathbf{v}}{dt} = -\nabla p + \mathbf{J} \times \mathbf{B}. \tag{1.2}$$

The perpendicular component of the current density is thus

$$\mathbf{J}_\perp = \frac{\mathbf{B} \times \nabla p}{B^2} - \frac{\rho \, d\mathbf{v}}{B^2 \, dt} \times \mathbf{B}, \tag{1.3}$$

if the pressure is isotropic.

The current generally has a divergence in the equatorial plane of the magneto-sphere, which must be compensated for by field-aligned currents, $j_\parallel$, flowing into and out of the ionosphere. This current continuity condition, which must be satisfied even if the convection is highly non-stationary, leads to the equation (Hasegawa and Sato 1979):

$$B\frac{\partial}{\partial l}\left(\frac{j_\parallel}{B}\right) + \nabla_\perp \cdot \mathbf{J}_\perp = 0, \tag{1.4}$$

where l is taken along the magnetic field line. Equivalently, one may write

$$\nabla_\perp \cdot \mathbf{J}_\perp = -\rho\frac{d}{dt}\left(\frac{\Omega}{B}\right) - \frac{\mathbf{J}_\perp \cdot \nabla B}{B^2} + \mathbf{J}_{in}\cdot\frac{\nabla n}{n}, \tag{1.5}$$

where

$$(\text{vorticity}) \quad \Omega = \nabla \times \mathbf{v}$$

$$(\text{inertia current}) \quad \mathbf{J}_{in} = \mathbf{B} \times \left(\frac{\rho}{B^2}\frac{d\mathbf{v}}{dt}\right)$$

$$(\text{mass density}) \quad \rho = mn.$$

The first term in Eq. (1.5) indicates the effect of dynamic changes in the vorticity and magnetic flux density. The second term denotes the effect of the spatial variation of the magnetic field in the direction of the perpendicular current. The source of this current may be called a current generator because its functional form depends only on the spatial structure, not on its dynamic changes. The current from the third term originates from density inhomogeneities in the direction of the inertia current.

Under the slow-flow approximation ($d\mathbf{v}/dt = 0$) Eq. (1.4) becomes

$$\frac{\partial}{\partial l}\left(\frac{j_\parallel}{B}\right) = -\frac{2(\mathbf{B} \times \nabla B)\cdot \nabla p}{B^4}. \tag{1.6}$$

Integrating this equation along the field line, one can reach

$$j_{\parallel i} = \frac{B_i}{2B_e}\left(\nabla p \times \frac{\mathbf{B}_e}{B_e}\right)\cdot\nabla\int\frac{dl}{B}, \tag{1.7}$$

where $\mathbf{B}_e$ is the magnetic field at the equator, B_i is the field magnitude at the ionosphere, and $j_{\parallel i}$ is the field-aligned current density per unit area in the ionosphere (positive for downward current). As will be discussed in Chapter 3, the contribution of the magnetospheric current to the field-aligned current can be expressed as the divergence of a magnetospheric "Hall" current:

$$j_\parallel \propto -\nabla\cdot\Sigma^*\mathbf{B} \times \mathbf{E}/B,$$

where the equivalent "Hall" conductivity Σ^* is given by $\Sigma^* = \int dl(ne/B)$ (Vasyliunas, 1972).

On the other hand, in the ionosphere these field-aligned currents must be balanced by the divergence of ionospheric conduction currents. These "three-dimensional currents" require a specific distribution of the space charge and ionospheric electric field, which can be computed from the given field-aligned currents. Namely,

$$\nabla \cdot \mathbf{J}_i = \nabla \cdot [\Sigma] \mathbf{E}_i = j_{\|i} \sin \chi, \tag{1.8}$$

where $\mathbf{J}_i$ is the ionospheric current, $[\Sigma]$ is the ionospheric conductivity tensor, and χ is the inclination of the magnetic field. To close the logical loop of Fig. 1.8, the ionospheric electric field must, in turn, be consistent with the driving magnetospheric electric field obeying generalized Ohm's law.

This entire treatment of the magnetosphere–ionosphere system should provide a reasonable model of electrodynamic processes associated with plasma convection, as far as the closed field line region of the magnetosphere is concerned. In the open field line region, however, plasma as well as field behavior must be closely related to interplanetary and boundary layer processes (see Sect. 1.1.1). On these open magnetic field lines, the electrodynamic quantities in the ionosphere can also be determined from Eq. (1.8), since it is still applicable if the field-aligned current is given.

It is interesting to note that the contribution of magnetospheric electrons to the total plasma pressure is typically an order of magnitude smaller than the ion contribution (e.g., Baumjohann et al. 1989), although the dominant feature of their dynamic behavior is their sensitivity to a variety of mechanisms by which they are precipitated into the polar ionosphere, heating the ionosphere and producing the auroral luminosity (Fontaine et al. 1985). While magnetospheric ions are responsible for the generation of field-aligned currents, electrons carry the field-aligned currents and their precipitation tends to modify the distribution of the ionospheric conductivities through which these field-aligned currents must close. This implies that both electrons and ions contribute to the determination of the exact distribution of horizontal electric fields in the ionosphere, which, when mapped back to the magnetosphere along field lines, determine the motion and distribution of magnetospheric particles.

2 Morphology of Electric Fields and Currents at High Latitudes

Following Maxwell's law, the distortion of the terrestrial dipole field into the typical magnetospheric shape is accompanied by electrical currents. As schematically shown in Fig. 2.1, the compression of the terrestrial magnetic field on the dayside is caused by currents flowing perpendicular to the dipole field lines across the magnetopause surface. These currents are often called Chapman–Ferraro currents since Chapman and Ferraro were the first to postulate such a current system, some 60 years ago. The tail-like field of the nightside magnetosphere is accompanied by the tail current flowing on the tail surface and the neutral sheet current in the central plasma sheet, both of which are connected and form a Θ-like current system, if seen from along the Earth-Sun line.

Another large-scale current system which influences the configuration of the inner magnetosphere is the ring current. This current encircles the Earth in a westward direction at radial distances of several R_E and is carried mainly by protons. The protons are trapped particles which have energies of some tens of keV, bouncing back and forth between their magnetic mirror points in the northern and southern hemispheres. Due to gradients in the plasma pressure and magnetic field, these protons experience a westward drift, leading to a westward electric current. The outer portions of the ring current merge with the tail current in the plasma sheet (Sugiura 1972).

In addition to these purely magnetospheric current circuits which all flow perpendicular to the ambient magnetic field, there is a another set of currents which flows along magnetic field lines. These field-aligned currents, which are often called Birkeland currents, connect the current systems in the magnetosphere and its boundaries to those flowing in the polar ionosphere. The field-aligned currents are essential for the exchange of energy and momentum between these regions.

2.1 Large-Scale Current Systems

There are quite a number of high-latitude current systems which can coexist at auroral latitudes and in the polar cap, depending on the state of the magnetosphere, whether quiet or disturbed, and of the solar wind, especially the direction of the interplanetary magnetic field. Some of these current circuits are global in nature, involving the whole polar ionosphere or a substantial part of it. Most notable are the convection electrojets, the substorm current wedge and the polar cap and cusp currents. The morphology and electrodynamics of these large-scale current systems will be the subject of this section. Small-scale current systems associated with various auroral forms will be discussed in Section 2.2.

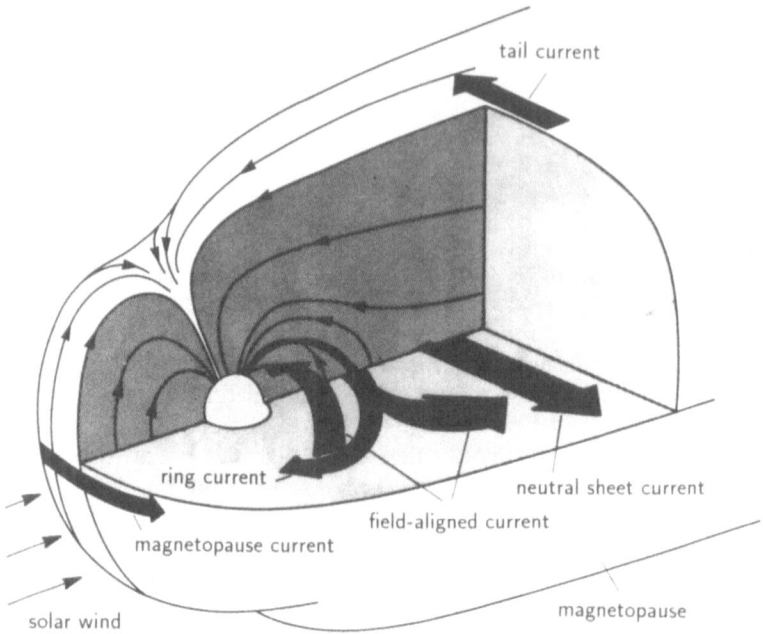

Fig. 2.1. Overview of current circuits flowing in the magnetosphere and on the magnetopause. (After Baumjohann and Haerendel 1987)

2.1.1 Convection Electrojets

The most prominent current system at auroral latitudes is the global convection electrojet system, carrying a total current of the order of 10^6 A. Our present knowledge on conductivity structure, electric fields, and current flow associated with the convection auroral electrojet system is summarized in Fig. 2.2. The auroral electrojets flow throughout the whole auroral oval, which is, in fact, an off-center ring shifted by an average 4° from the invariant magnetic pole toward magnetic midnight (Meng et al. 1977).

Inside the auroral oval the ionospheric conductivity is enhanced above the solar UV-induced level due to the ionization of neutral atoms and molecules by precipitating electrons and, to a lesser extent, ions. The energetic particles drift toward and around the Earth under the combined effect of $E \times B$, gradient, and curvature drift and precipitate, depending on their energy and pitch angle, in different local time sectors (e.g., McDiarmid et al. 1975; Hardy et al. 1985). The number of precipitating particles can be enhanced by pitch angle diffusion or scattering due to plasma waves, which brings more particles into the loss cone (e.g., Lyons 1983). The precipitation pattern is reflected in the conductivity structure, depending on the energy of the precipitating particles. The weakest conductivities are found near the noon sector, while the conductivity maximum lies in the midnight sector where typical values of 7–10 and 10–20 S for Pedersen

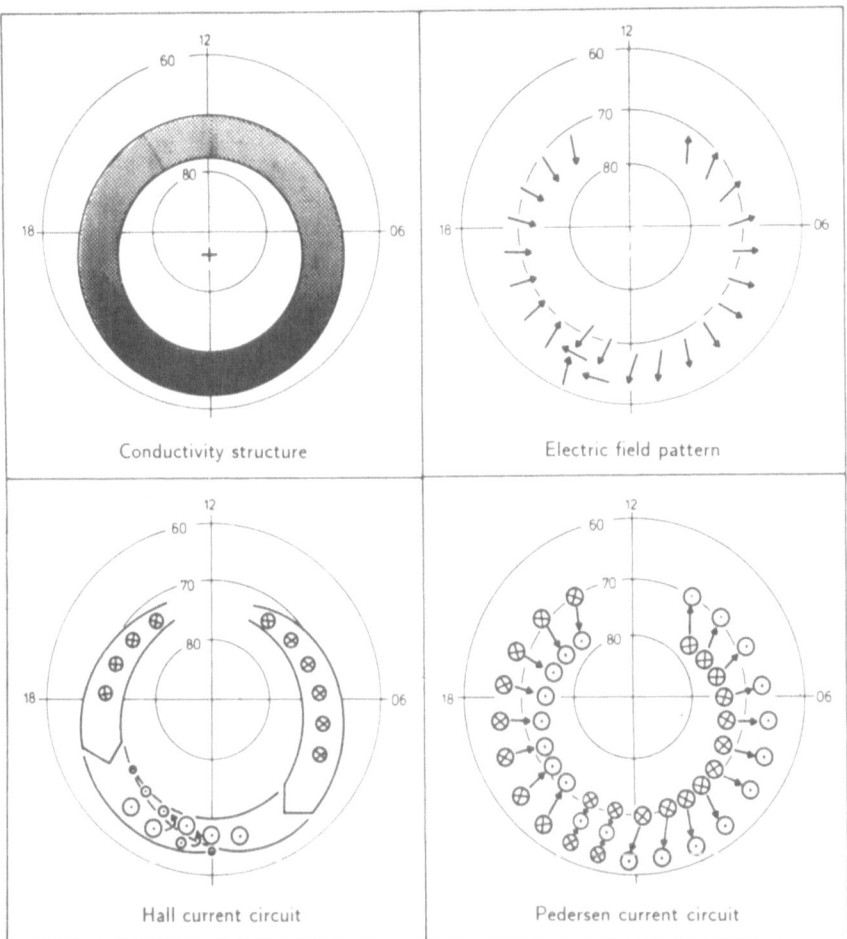

Fig. 2.2. Schematic diagram of conductivity structure, electric field pattern (direction only), and current flow within the convection auroral electrojet system. Coordinates are invariant latitude and magnetic local time

and Hall conductances (height-integrated conductivities), respectively, were found by, for example, Wallis and Budzinski (1981).

The electric field pattern in the auroral oval reflects the large-scale pattern of magnetospheric plasma convection associated with solar wind–magnetosphere coupling (cf. Sect. 1.1.1). The transport of open and closed flux tubes, depicted in Fig. 1.3, results in the convection pattern shown in Fig. 2.3. The electric field associated with this two-cell system of plasma transport has typical values between 20 mV/m during quiet times and 50 mV/m under active geomagnetic conditions at auroral latitudes. It is poleward directed in the afternoon and early evening sector, points equatorward in the postmidnight and morning sector, and rotates from north over west to south in the premidnight sector (e.g., Foster

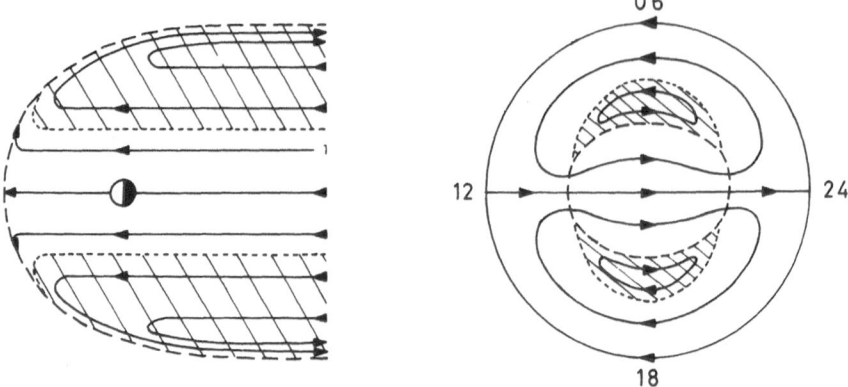

Fig. 2.3. Illustration of large-scale magnetospheric convection in the equatorial plane (*left*) and the polar ionosphere (*right*). The *hatched region* indicates convection driven by viscous-like interaction, while the remainder of the flow is associated with the reconnection process. (Cowley 1982)

et al. 1981; Zi and Nielsen 1982). This region of field rotation is called the Harang discontinuity region (cf. Happner 1972a; Maynard 1974; Kamide 1978). The Harang discontinuity is the locus of points at nighside auroral latitudes across which the meridional component of the ionospheric electric field is reversed from a poleward field in the eastward electrojet region to an equatorward field in the westward electrojet region (see also Heelis and Hanson 1980; Atkinson 1984; Erickson et al. 1991).

In contrast to the conductivities and electric fields, it is difficult to directly measure the ionospheric and field-aligned currents. Thus, the intensity and the distribution of these currents have to be estimated from ground-based and/or satellite measurements of magnetic fields using the methods described in Sections 4.1 and 4.2. Numerous studies of this kind have been made in the past (e.g., Zmuda and Armstrong 1974a,b; Iijima and Potemra 1976; Hughes and Rostoker 1979; Baumjohann et al. 1980; Kamide et al. 1981) and yielded fairly consistent results, as far as the average picture is concerned. They are summarized in the two lower panels of Fig. 2.2.

Both eastward and westward electrojets are primarily Hall currents which originate around noon and are fed by downward net field-aligned currents, i.e., field-aligned currents which are not balanced along a meridian. Their sheet current densities range between 0.5 and 1 A/m and increase toward midnight due to the increasing Hall conductance. The eastward electrojet terminates in the region of the Harang discontinuity where it partially flows up magnetic field lines and partially rotates northward, joining the westward electrojet. This eastward current originates in the pre-noon sector, flows through the morning and midnight sector, and typically extends into the evening sector along the poleward border of the auroral oval where it also diverges as upward field-aligned currents.

The Pedersen current with typical densities of 0.3–0.5 A/m flows northward in the eastward electrojet region and is connected to sheets of downward and upward field-aligned currents in the southern and northern half of the afternoon-evening northern hemisphere auroral oval, respectively. In the midnight-to-noon sector the Pedersen current flows equatorward and field-aligned currents provide continuity by flowing upward in the southern and downward in the poleward half of the auroral oval. Iijima and Potemra (1976) named the field-aligned currents in the poleward half of the auroral oval Region 1 currents, and those in the equatorward half Region 2 currents.

The Pedersen current circuit in the Harang discontinuity was a matter of debate for some years. For example, from satellite measurements Iijima and Potemra (1987) deduced three field-aligned current sheets in this local time sector, while Kamide (1978) proposed a model for the current flow in the Harang discontinuity region which contained only two field-aligned current sheets. The

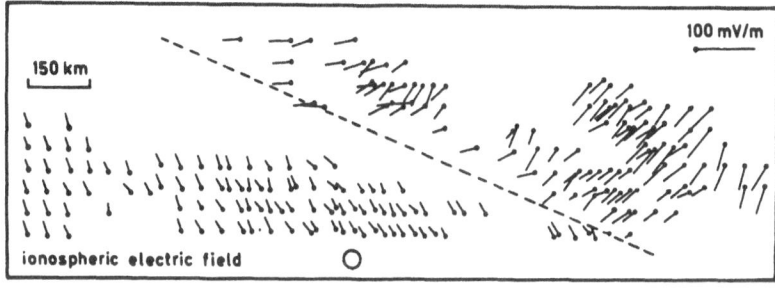

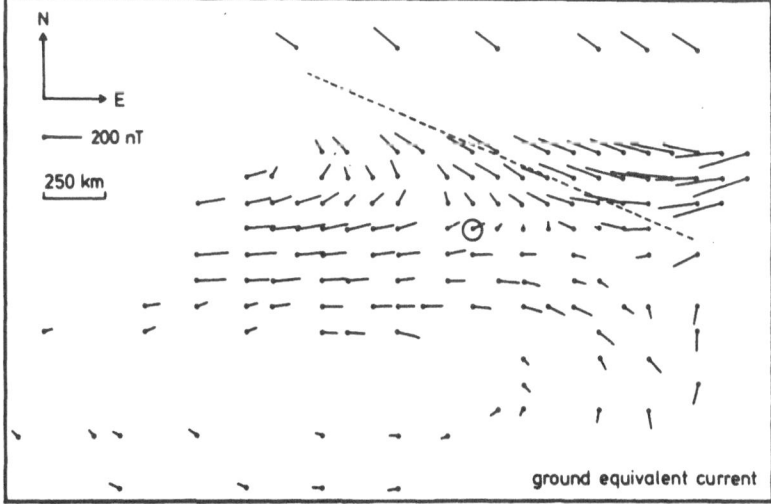

Fig. 2.4. Electric field and equivalent current vectors in the vicinity of the Harang discontinuity. Both panels are constructed by superimposing several two-dimensional vector distributions measured during a westward passage of the Harang discontinuity over northern Scandinavia. The *large, open circles* serve as reference points between the two panels (and those of Fig. 2.5). The *dashed lines* give the (electrical) Harang discontinuity. (After Kunkel et al. 1986)

more refined measurements used by Robinson et al. (1985a) and Kunkel et al. (1986) were needed to resolve this controversy.

Figure 2.4 summarizes data of the electric and magnetic field in two dimensions measured simultaneously by the STARE coherent backscatter radar system (Greenwald et al. 1978) and the Scandinavian Magnetometer Array (Küppers et al. 1979) during the westward passage of the Harang discontinuity over norther Scandinavia (Kunkel et al. 1986). The electric field panel displays northward electric fields which gradually rotate with increasing latitude through the westward (in the Harang discontinuity region) toward a southwesterly direction. The equivalent current panel contains an eastward electrojet which diverges southward as well as northward in the eastern half of the observation area. The northward equivalent current joins the westward electrojet intruding from the east while the southward equivalent current forms an equivalent return "current" loop at subauroral latitudes.

Figure 2.5 gives the electrodynamic parameters of the best-fit model current system. These parameters where deduced by Kunkel et al. (1986) by complementing the pattern of the interpolated electric field with a first-guess conductance distribution, calculating ionospheric and field-aligned currents and associated ground magnetic fields, and then subsequently changing the conductance distribution until the model fields fit the measured magnetic field pattern as closely as possible.

The deduced conductance distribution exhibits two separate bands in the eastward and westward electrojet region with relatively low Hall conductances in the vicinity of the Harang discontinuity. Since the Pedersen conductance does not decrease so strongly there, this results in a minimum of the Hall-to-Pedersen conductivity ratio co-located with the Harang discontinuity.

The field-aligned current panel shows upward current in and near the Harang discontinuity, in agreement with Robinson et al. (1985a). The upward field-aligned current just inside the discontinuity stems from Pedersen currents flowing in from both the north and the south. The upward currents located a bit further to the south and the north are fed by the Hall current electrojets.

2.1.2 Substorm Current Wedge

During magnetospheric substorms, the ionospheric current flow is affected in two ways (e.g., Baumjohann 1983, 1986; Rostoker et al. 1987a) to accommodate the enhanced energy input from the solar wind via increased dissipation. On the one hand, the Hall current flow in the convection auroral electrojets described in Section 2.1.1 (the so-called DP2 system; see the left panel of Fig. 2.6) increases in direct relation to the energy input from the solar wind along with the driven process. Additionally, sporadic unloading of energy previously stored in the magnetotail leads to the formation of a substorm current wedge (the so-called DP1 system; see right panel of Fig. 2.6) with strongly enhanced westward current flow in the midnight sector.

During a substorm, the solar wind energy which is not directly dissipated via enhanced convection in the auroral electrojets is stored in the tail magnetosphere,

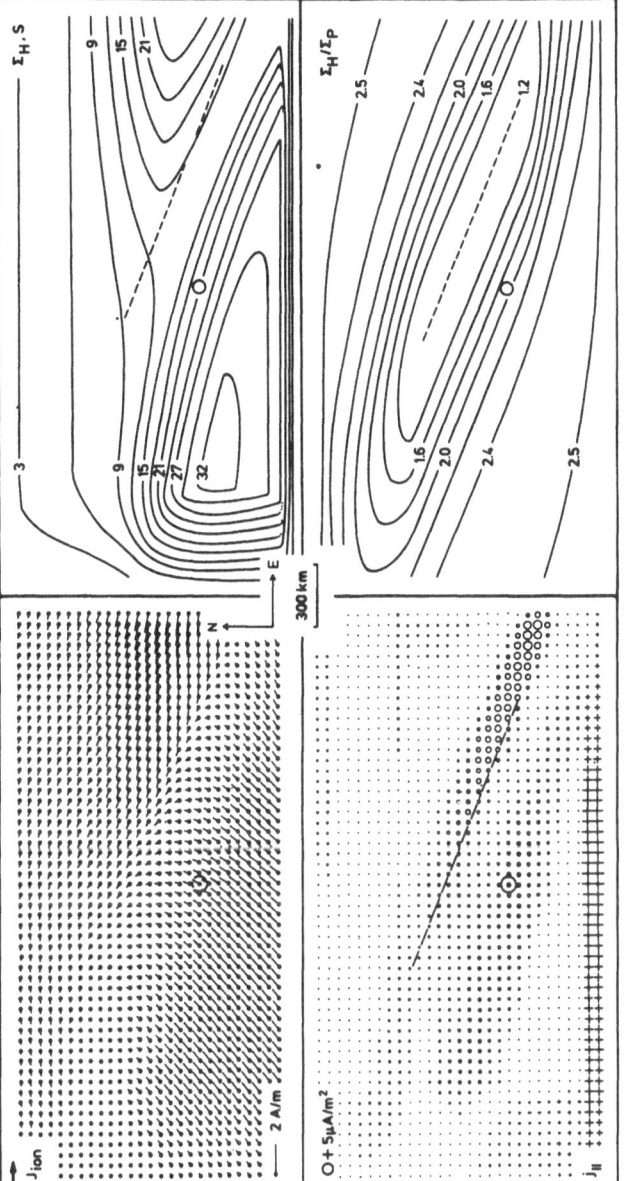

Fig. 2.5. Two-dimensional distribution of ionospheric current (*upper left*), Hall conductance (*upper right*), field-aligned current density (*lower left*), and Hall-to-Pedersen conductivity ratio (*lower right*) for the best-fit model reproducing the electric field and equivalent current measurements of Fig. 2.4. *Circles and crosses* denote upward and downward directed current flow; *large, open circles and dashed lines* bear the same meaning as in Fig. 2.4. (After Kunkel et al. 1986)

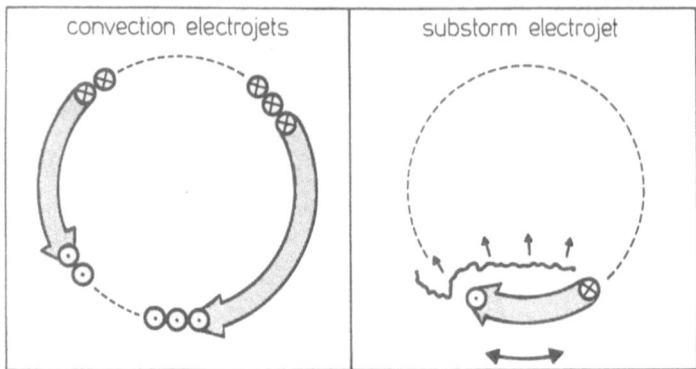

Fig. 2.6. Illustration describing location, flow direction and field-aligned current closure of the convection electrojets (DP 2) and the substorm current wedge (DP 1). (Baumjohann 1983)

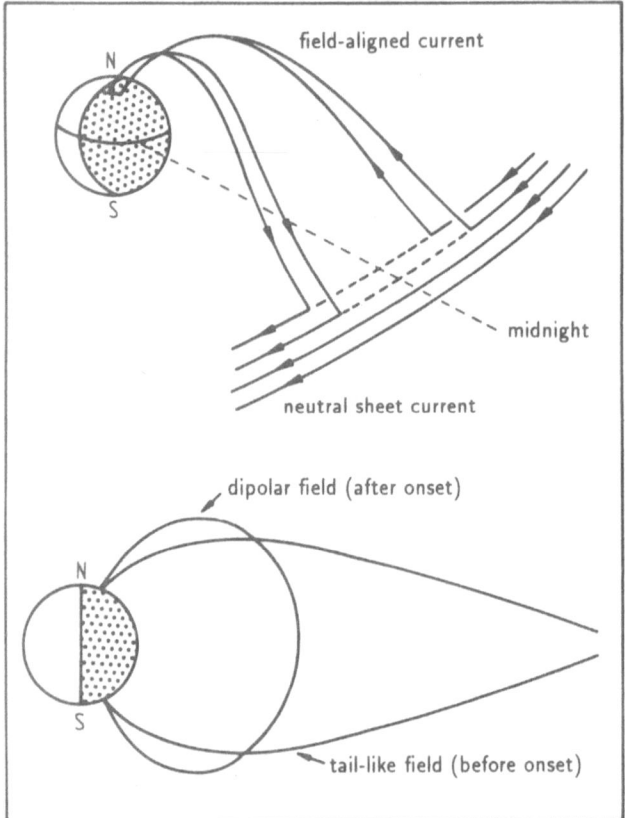

Fig. 2.7. *Above*, The substorm current wedge short-circuits the neutral sheet current in the magnetotail. *Below*, Reconfiguration of the nightside magnetospheric magnetic field associated with the disruption of the cross-tail current

One of the conclusions from the computer simulations for two sets of models for $\Sigma_G = 100$ S (simulating a voltage generator) and 0.1 S (simulating a current generator), is that the scale size is smaller for current generators, indicating that the narrow discrete arcs may be current-driven while the large-scale auroral currents are voltage-driven.

A voltage source is, by definition, an infinite energy source of zero internal resistance. Under such conditions the electric potential across the source region can be shown to stay constant regardless of how the load, i.e. the ionosphere, changes with time. From the wave propagation standpoint, this may mean that

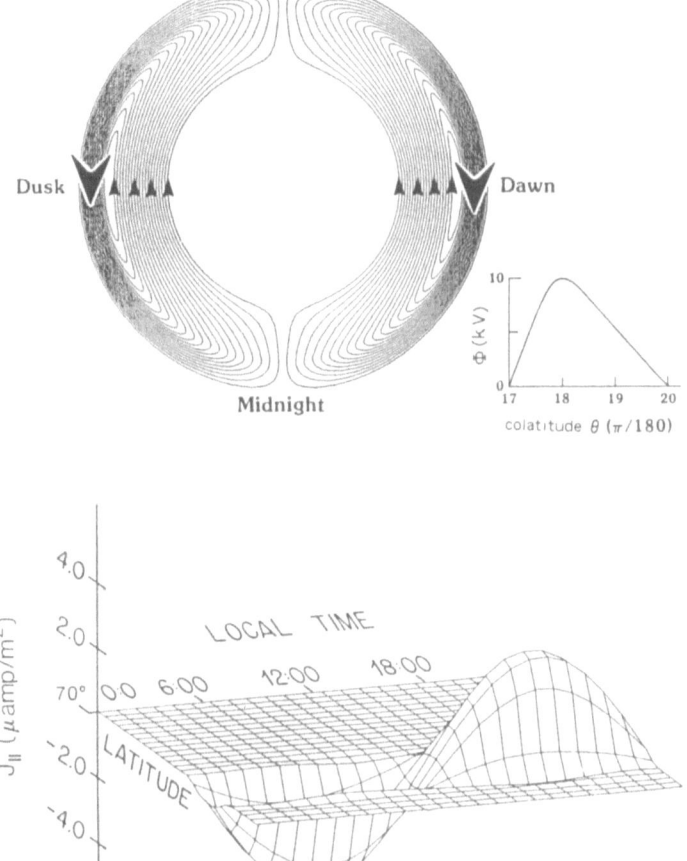

Fig. 3.9. b *Top,* Equipotential contours of the electrostatic potential in the ionosphere (or plasma flow pattern in the magnetospheric equatorial plane) given as the source of the magnetosphere–ionosphere coupling and the potential from $\Phi\,(\theta)$. *Bottom,* Perspective view of the field-aligned current distribution at the final stage of the time-dependent simulation model. (Watanabe et al. 1986).

incident and reflected wave fields cancel each other: the nature of more general configurations is discussed by Glaßmeier (1984). In reality, since the solar wind does contain a large amount of plasma flow energy, the source region of open field lines may be approximated by a constant-voltage generator. On the other hand, on closed field lines in the plasma sheet, the source region cannot be expected to behave like a voltage generator. If this consideration is correct, whether the region is a voltage or current source depends primarily on whether the region is open or closed.

On the other hand, the reflection at the ionosphere of the Alfvén waves carrying changes in the field-aligned currents depends strongly on the ionospheric conductivity. The ionospheric conductivity itself is not independent of the field-aligned currents. For example, it increases in the region of upward field-aligned currents where energetic electrons are precipitating. This leads to a feedback process which can result in "traveling" conductivity enhancements. Lysak (1986) studied the unstable nature of this feedback.

In the second extensive study, Watanabe et al. (1986) conducted three-dimensional modeling of the magnetosphere–ionosphere coupling, where one-fluid but time-dependent MHD equations are solved. In their scheme, a simple model of the electric field is initially given in the magnetospheric equatorial plane (see the top of Fig. 3.9b). It is quite reasonable to assume that such a plasma flow in the equatorial plane is sustained by the solar wind–magnetosphere interaction (see Chap. 1). For simplicity, the ionospheric conductivity is uniform and unchanged.

The practical procedure of this modeling is that (1) for a given electric field, the field-aligned current is determined; (2) given the field-aligned current distribution, the ionospheric electric field is calculated; and (3) going back to the magnetosphere, magnetospheric equations are solved to advance the time integration. The parallel electric field is disregarded. This procedure is equivalent to the situation where Alfvén waves propagate down to the ionosphere, accompanying the field-aligned current and exciting the electrostatic potential in the ionosphere.

After some time within the modeling scheme, the ionosphere responds to the given electric field and reaches some kind of steady state, if the original field is not changed. The time scale of the potential evolution depends on the ratio of the ionospheric conductivity and the corresponding quantity in the magnetosphere. For example, the ionospheric potential grows suddenly at $t = t_0$ (the Alfvén transit time) when the wave reaches the ionosphere. Figure 3.9a shows how the electrostatic potential in the ionosphere responds to two cases of the ionospheric plasma density, $N_0 = 1.25 \times 10^{10}/m^3$, and $2.5 \times 10^{10}/m^3$ respectively. The time development of the ionospheric response for these two cases, corresponding to the typical nightside and dayside ionosphere, indicates that the ionospheric potential grows in a different manner, depending on different values of the ionospheric conductivity. Shown at the bottom of Fig. 3.9b is the resultant field-aligned current distribution at the final stage. As expected, the potential contours (not shown here) are similar to the assumed plasma flow pattern. The field-aligned current calculated from this modeling reproduces the main features that are indicated by satellite observations. Although it is as yet preliminary, in the sense that they solved MHD equations for zero parallel potential and constant ionospheric conductivity, the result still confirms many observed features during disturbed times, including that the typical twin-vortex potential pattern can develop slowly or faster, depending on the conductivity.

Watanabe and Sato (1988) further included time changes in the ionospheric density: the ionosphere is no longer a stationary and passive medium for information coming from the magnetosphere, but acts as an active medium. The feedback coupling between Alfvén waves of magnetospheric origin and ionospheric density waves results in the reproduction of "spontaneous" enhancements of field-aligned currents and the ionospheric density, and thus auroral arcs.

The third model is by Zi and Shen (1986), who developed a time-dependent, analytical model of the ionospheric consequence of magnetospheric convection. For this purpose, the effect of the day-night gradients in the ionospheric conductivity on the efficiency of magnetosphere–ionosphere coupling is included. For a relatively simple driving potential, which is assumed to be applied initially to the magnetosphere and to remain invariant later on, the time evolution of the global potential distribution has been calculated. The distribution of the field-aligned currents predicted by the model calculation is in agreement with the statistical configuration of recent satellite measurements. It is also found that the time constant of the evolution toward the steady state varies with local time, depending on the non-uniformity of the ionospheric conductivity.

4 Modeling of Ionospheric Electrodynamics

The self-consistent chain for magnetosphere–ionosphere coupling processes represents a closed loop, the entire chain consisting of a number of individual links. Studies of theoretical models, which intend to examine the entire loop, and those attempting to understand "local" chains, or individual links, must be complementary to each other. Sato (1985) called the latter "local" key regions, where remarkable energy conversion processes are taking place. The purpose of this chapter is to present the basic equations for simple models with special emphasis placed on ionospheric electrodynamic processes that result from the polar ionosphere and field-aligned currents. We attempt to obtain a set of numerical solutions of the equations governing the magnetosphere–ionosphere system in such a way as to include far less idealization or simplification by referring to more essential observational characteristics than would be required for pure theories. In this way, it is possible to demonstrate how the basic assumptions lead to the main observed signatures in the high-latitude ionosphere, where many exciting phenomena occur.

4.1 Ionospheric Parameters Controlled by Field-Aligned Currents

4.1.1 Basic Algorithm

Described in this subsection are the basic equations that are commonly used in the numerical calculations of the electric potential, field, and current in the high-latitude ionosphere. In order to simplify conditions, a number of assumptions are made throughout the entire calculation process. The most important and crucial ones are as follows: (1) the ionosphere is regarded as a two-dimensional spherical current sheet with a height-integrated layer conductivity, since we are interested only in the large-scale current and field patterns involving distances much longer than the thickness of current layers within the altitude range of the ionosphere. (2) The Earth's magnetic field lines are taken to be equipotentials, neglecting parallel electric fields. (3) Only steady-state solutions are considered.

The continuity equation for electric currents under these simplifying conditions is given by

$$\operatorname{div} \mathbf{J} = j_{\parallel} \sin\chi, \tag{4.1}$$

where $\mathbf{J}$ is the height-integrated ionospheric current density, $j_{\parallel}$ is the density of the field-aligned current (positive for a downward current), and χ is the inclination angle of a geomagnetic field line with respect to the horizontal ionosphere.

Ohm's law for the ionospheric current is written as

$$\mathbf{J} = (\Sigma)\mathbf{E} = -(\Sigma)\cdot\text{grad }\Phi, \tag{4.2}$$

where $\mathbf{E}$ and Φ are the electric field and the corresponding electrostatic potential, respectively, in the frame rotating with the Earth, and (Σ) is the dyadic of the height-integrated ionospheric conductivity. The combination of Eqs. (4.1) and (4.2) yields

$$\text{div}\,[(\Sigma)\cdot\text{grad }\Phi] = -j_{\|}\sin\chi, \tag{4.3}$$

which is to be solved for Φ under suitable boundary conditions. The boundary conditions we employ can be:

$$\Phi = 0 \quad \text{(at the poles)} \tag{4.4a}$$

$$\partial\Phi/\partial\theta = 0 \quad \text{(at the equator)}, \tag{4.4b}$$

where θ is the colatitude. As long as we concentrate on the high-latitude (say, $\theta < 30°$) ionosphere, the boundary conditions are not very important. In solstitial seasons, however, an asymmetry in the ionospheric conductivity between the northern and southern hemispheres is present, and hence the resultant electric currents are also expected to be asymmetric. In such cases, actual calculations are made for both the northern and southern hemispheres, but the results are shown only for one hemisphere, depending on the season.

Given a specific conductivity model and an assumption about the field-aligned current distribution, one can solve Eq. (4.3) to find the electric potential Φ in the high-latitude ionosphere. Note that by regarding the ionosphere as a two-dimensional thin layer (see the above, simplifying assumption), this modeling scheme cannot be applied directly to the equatorial region, where the field lines are nearly parallel to the Earth's surface, field-aligned currents hence being horizontal ionospheric currents there. The differential equation corresponding to Eq. (4.3) is then solved numerically to obtain the most probable potential value at each of the grid points in the scheme.

By using the (θ, λ) coordinate system, in which θ is colatitude and λ is longitude, measured eastward from midnight, Eq. (4.3) can be reduced to the form

$$A\frac{\partial^2\Phi}{\partial\theta^2} + B\frac{\partial\Phi}{\partial\theta} + C\frac{\partial^2\Phi}{\partial\lambda^2} + D\frac{\partial\Phi}{\partial\lambda} = F,$$

where

$$A = \sin^2\theta\Sigma_{\theta\theta}$$

$$B = \sin\theta\left[\frac{\partial}{\partial\theta}(\sin\theta\Sigma_{\theta\theta}) - \frac{\partial\Sigma_{\theta\lambda}}{\partial\lambda}\right] \tag{4.5}$$

$$C = \Sigma_{\lambda\lambda}$$

$$D = \sin\theta\frac{\partial\Sigma_{\theta\lambda}}{\partial\theta} + \frac{\partial\Sigma_{\lambda\lambda}}{\partial\lambda}$$

$$F = -a^2 j_\parallel \sin^2\theta \sin\chi.$$

$$\sin\chi = \frac{2\cos\theta}{(1 + 3\cos^2\theta)^{1/2}}$$

and a is the radius of the current sheet. Note that if all the conductivity gradients are neglected Eq. (4.5) is reduced to the Poisson equation. Further, in the region where $j_\parallel = 0$ the Laplace equation is obtained.

The component of the height-integrated ionospheric current is given by

$$\begin{bmatrix} J_\theta \\ J_\lambda \end{bmatrix} = \begin{bmatrix} \Sigma_{\theta\theta} & \Sigma_{\theta\lambda} \\ -\Sigma_{\theta\lambda} & \Sigma_{\lambda\lambda} \end{bmatrix} \begin{bmatrix} E_\theta \\ E_\lambda \end{bmatrix}, \tag{4.6}$$

where the electric field component is expressed as

$$E_\theta = -\frac{\partial\Phi}{a\partial\theta} \quad E_\lambda = -\frac{\partial\Phi}{a\sin\theta\,\partial\lambda}. \tag{4.7}$$

It is also worth computing the equivalent ionospheric current systems, in order to compare them with the overhead equivalent current systems estimated directly from ground-based magnetic observations, which include both ionospheric and field-aligned current effects. As is done by Vasyliunas (1970a), we can separate the ionospheric current **J** into two elements: first $\mathbf{J}_T$ (source-free ionospheric current or the toroidal current), which is confined within the ionosphere, and, second $\mathbf{J}_P$ (closing current via $j_\parallel$ or a "potential" current), which serves merely to close the field-aligned current.

Thus, we have

$$\mathbf{J} = \mathbf{J}_T + \mathbf{J}_P, \tag{4.8}$$

where, by definition,

$$\operatorname{div}\mathbf{J}_T = 0, \quad \operatorname{curl}\mathbf{J}_P = 0. \tag{4.9}$$

On the assumption that $\mathbf{J}_P$ can be derived from the associated scalar function as

$$\mathbf{J}_P = \operatorname{grad}\tau, \tag{4.10}$$

the function τ can be calculated from the following Poisson equation:

$$\frac{1}{a^2\sin\theta}\frac{\partial}{\partial\theta}\left(\sin\theta\frac{\partial\tau}{\partial\theta}\right) + \frac{1}{a^2\sin^2\theta}\frac{\partial^2\tau}{\partial\lambda^2} = -j_\parallel \sin\chi. \tag{4.11}$$

Once τ is solved, $\mathbf{J}_P$ and $\mathbf{J}_T$ can easily be obtained from Eq. (4.10). Note that in the absence of conductivity gradients, $\mathbf{J}_T$ is simply the Hall current and $\mathbf{J}_P$ is the Pedersen current.

As suggested by Fukushima (1969, 1976) and Vasyliunas (1970b) for uniform conductivity conditions, it can be assumed that the ground-level magnetic field caused by $j_\parallel$ is the same as that caused by $-\mathbf{J}_P$. The equivalent electric current system estimated from ground observations for the magnetic field produced by **J** and $j_\parallel$ (hence, **J** and $-\mathbf{J}_P$) is identical to $\mathbf{J}_T$.

We can also obtain the equivalent current system by drawing isointensity contours of the corresponding current function. One may introduce the current function ψ, defined by

$$\mathbf{J}_T = -\operatorname{grad}\psi \times \mathbf{n}_r, \tag{4.12}$$

where $\mathbf{n}_r$ denotes the unit vector in the radial direction from the center of the Earth. Contours of $\psi = \text{const}$, then, give streamlines of the equivalent ionospheric current, $\mathbf{J}_T$. By combining Eqs. (4.2) and (4.12), the relationship between the electric potential and the current function can be deduced as

$$\frac{1}{a\sin\theta}\frac{\partial\psi}{\partial\lambda} = -\Sigma_{\theta\theta}\frac{\partial\Phi}{a\partial\theta} - \frac{\Sigma_{\theta\lambda}}{a\sin\theta}\frac{\partial\Phi}{\partial\theta} + \frac{\partial\tau}{a\partial\theta}, \tag{4.13a}$$

$$\frac{\partial\psi}{a\partial\theta} = \Sigma_{\theta\lambda}\frac{\partial\Phi}{a\partial\theta} - \frac{\Sigma_{\lambda\lambda}}{a\sin\theta}\frac{\partial\Phi}{\partial\lambda} + \frac{1}{a\sin\theta}\frac{\partial\tau}{\partial\lambda}. \tag{4.13b}$$

4.1.2 Quiet Periods

Figure 4.1 shows a schematic representation of the simulation scheme used in a series of numercial studies by Kamide and Matsushita (1979a,b). Two representative cases during quiet periods are chosen for demonstration from their simulation studies. These are to simulate (a) a very quiet period without auroral enhancement, and (b) quiet periods with weak auroral enhancements. The maximum field-aligned current, whose density is $j_{\|0}$, flows into the ionosphere are the 0600 LT meridian in Fig. 4.1 and flows out from the 1800 LT meridian for all cases. The amount of $j_{\|0}$ is assumed to be $1.0 \times 10^{-7}\,\text{A/m}^2$ for (a) and 2.0×10^{-7} A/m^2 for (b). The intensity of the positive (downward) field-aligned current is distributed between colatitudes of $20°$–$30°$ and longitudes of $0°$–$180°$.

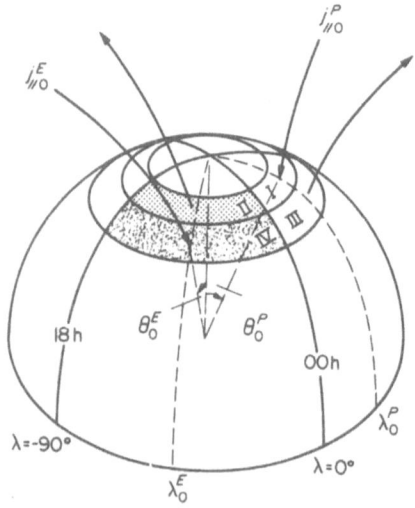

Fig. 4.1. Illustration of modeling of the polar ionosphere, in which field-aligned currents and ionospheric conductivities are given as functions of colatitude and longitude

It is assumed that the field-aligned currents are continuously distributed in the Gaussian form

$$j_\parallel = \pm j_{\parallel 0} \exp\left[-\frac{(\theta - \theta_0)^2}{D_\theta^2} - \frac{(\lambda \mp \lambda_0)^2}{D_\lambda^2} \right] \tag{4.14}$$

in the latitudinal belt between $\theta = 20°$ and $\theta = 30°$. The downward field-aligned current is flowing into the morning half of the auroral ionosphere, whereas the upward current is flowing out of the ionosphere into the evening half. This configuration may represent that of an extremely quiet time. The parameters, D_θ and D_λ, are chosen in such a way that the intensity of $j_\parallel$ becomes approximately $0.2 j_{\parallel 0}$ at the boundaries. Thus, the total downward (or upward) field-aligned current $I_\parallel$ amounts to

$$I_\parallel = \int_{20°}^{30°} \int_{0°}^{180°} j_\parallel a^2 \sin\theta \, d\theta \, d\lambda = 1.9 \times 10^5 \, \text{A}. \tag{4.15}$$

a) Very quiet period without auroral enhancement
The effects of slowly varying ionospheric conductivity without auroral enhancements are examined by using a fairly realistic distribution of the height-integrated conductivity (Tarpley 1970; Richmond et al. 1976). The height-integrated conductivities are written as:

$$\Sigma_{\lambda\lambda}(\theta, \lambda) = \Sigma_1^*(\theta) \sin \chi \, f(\cos K)$$
$$\Sigma_{\theta\theta}(\theta, \lambda) = \Sigma_1^*(\theta)/\sin \chi \, f(\cos K) \tag{4.16}$$
$$\Sigma_{\theta\lambda}(\theta, \lambda) = \Sigma_2^*(\theta) f(\cos K),$$

where Σ_1^* and Σ_2^* are the magnetic-field-integrated Pedersen and Hall conductivities for an overhead sun, and $f(\cos K)$ is a function describing the decrease in height-integrated conductivity with increasing solar zenith angle K. Conductivities $\Sigma_{\theta\theta}$, $\Sigma_{\lambda\lambda}$ and $\Sigma_{\theta\lambda}$ correspond to the classical height-integrated layer conductivities Σ_{xy}, Σ_{yy} and Σ_{xy} (Matsushita 1967; Rishbeth and Garriott 1969). The solar zenith angle K is determined in polar coordinates (θ, λ) by

$$\cos K = \cos\theta \cos\theta_s + \sin\theta \sin\theta_s \cos(\lambda - \lambda_s), \tag{4.17}$$

where (θ_s, λ_s) are the subsolar point coordinates. Figure 4.2 shows the distribution of $\Sigma_{\theta\theta}$, $\Sigma_{\lambda\lambda}$ and $\Sigma_{\theta\lambda}$ along noon and midnight meridians for $\theta_s = 90°$ and $\lambda_s = 180°$ representing equinox.

Figure 4.3 shows the computed potential distribution. The location of both the highest and lowest potentials moves toward the midnight meridian from the centers of the downward and upward field-aligned currents. This is simply because the nightside conductivity is smaller than the dayside conductivity, so that the electric field should be larger in the nightside than in the dayside to hold the current continuity. There is considerable asymmetry in the polar cap electric field strength, which is larger in the early morning sector than in the evening sector.

It is rather surprising to see that this tendency of the field asymmetry is the same as that seen during periods of the IMF (interplanetary magnetic field) away

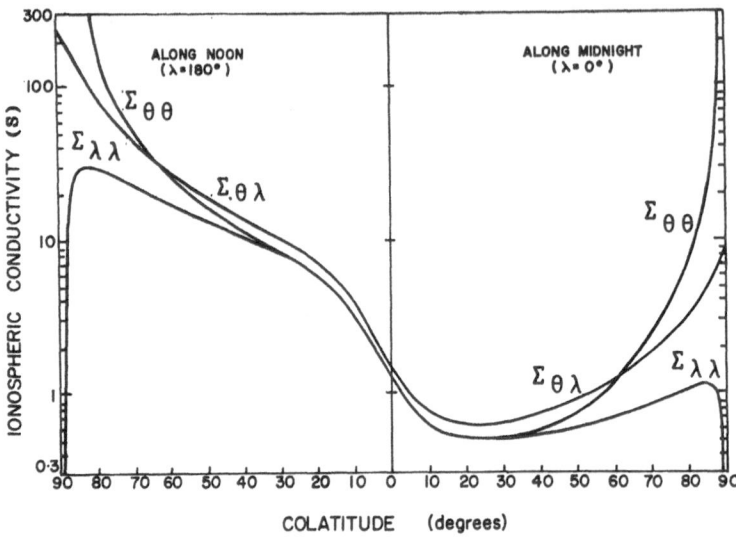

Fig. 4.2. Height-integrated conductivity distributions along the noon-midnight meridian in the equinoctial season. This conductivity model is used to simulate quiet-time conductivity distributions. For disturbed times, the effects of auroral enhancements are superimposed on this quiet-time model. (Kamide and Masushita 1979a)

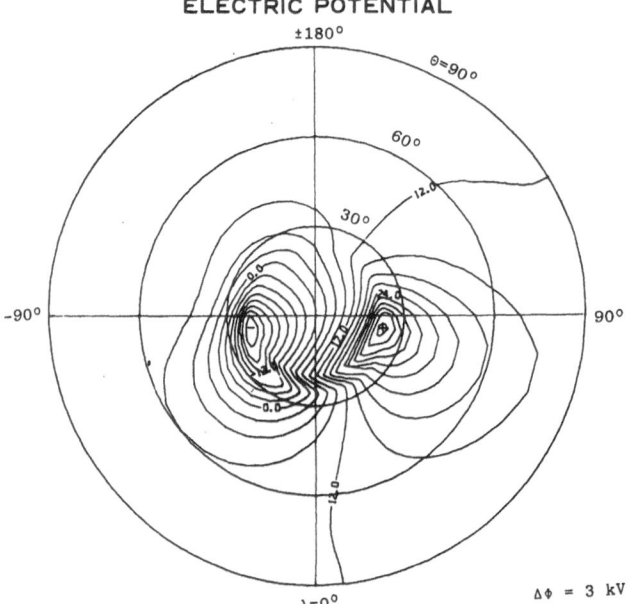

Fig. 4.3. Electric equipotential distribution (3-kV contour interval) over the northern hemisphere ($\theta = 90°$ is the equator and $\lambda = 0°$ is the midnight meridian) for the very quiet period without auroral enhancement. (Kamide and Matsushita 1979a)

sector (e.g. Heppner 1973). That is the asymmetry in the polar cap electric field can be reproduced without invoking changes in the IMF conditions. This problem has been discussed by Atkinson and Hutchinson (1978) for relatively simple models of the ionospheric conductivity in the polar cap. Accordingly, in order to be commensurate with the assumed symmetric field-aligned current distribution between the morning and evening sectors, the magnitude of the northward field in the evening auroral belt should be larger than that of the southward field in the morning sector. In fact, the large northward electric field in the evening sector (approximately 20 mV/m) has been observed by Chatanika incoherent scatter radar as a typical quiet-time feature (Horwitz et al. 1978).

b) Quiet cases with weak auroral enhancement

Even without substorm activity, visible auroras and auroral precipitation are known to be present along a continuous belt called the auroral oval (e.g. Lui et al. 1975). Winningham et al. (1975) have shown, using precipitating electron ($10\,\mathrm{eV} < E < 15\,\mathrm{keV}$) spectrograms, that the quiet-time auroral belt is characterized by diffuse auroras caused by low-energy electron precipitation. Data obtained from Chatanika incoherent scatter radar indicate that the height-integrated ionospheric conductivities are usually less than 5 S during quiet times (Brekke et al. 1974; Banks and Doupnik 1975; Horwitz et al. 1978).

To simulate such a quiet condition, a slight enhancement along the nightside auroral belt is included. The effects of a pair of field-aligned currents in each longitude are also considered. In each of the studies the following conductivity model is employed to simulate the quiet-time, diffuse, auroral enhancement:

$$\Sigma_{\theta\lambda} = \Sigma_{2m}\exp\left[-\frac{(\theta-\theta_0')^2}{(D_0)^2} - \frac{(\lambda-\lambda_0')^2}{(D_\lambda)^2} \right]$$
$$\Sigma_{\theta\theta} = \tfrac{1}{2}\Sigma_{\theta\lambda}/\sin\chi, \qquad\qquad\qquad\qquad\qquad (4.18)$$
$$\Sigma_{\lambda\lambda} = \tfrac{1}{2}\Sigma_{\theta\lambda}\sin\chi$$

where $\Sigma_{2m} = 10$ S.

Effects of Auroral Enhancement

Figure 4.4 shows the calculated potential distribution. An important point is that even though the field-aligned currents are assumed to be twice the value used in the extremely quiet case, the total potential difference across the polar cap does not increase very much, namely, from 54 kV (Fig. 4.3) to only 60 kV (Fig. 4.4). The most significant change, which is caused by an addition of the weakly conducting strip in the nightside auroral belt, is that although the electric field direction had originally been westward near midnight, it is now southwestward in the poleward half of the auroral belt and northwestward in the equatorward half. The corresponding ionospheric current vectors (not shown here) are characterized by the westward electrojet in the night sector and the eastward current at the dayside high latitudes.

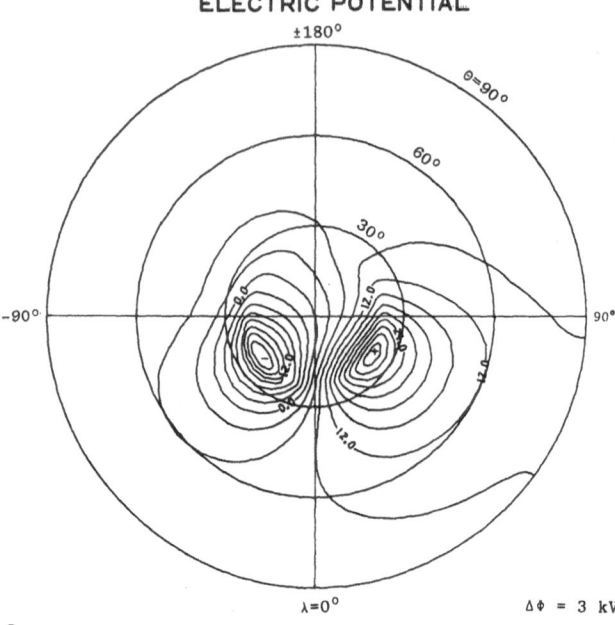

a

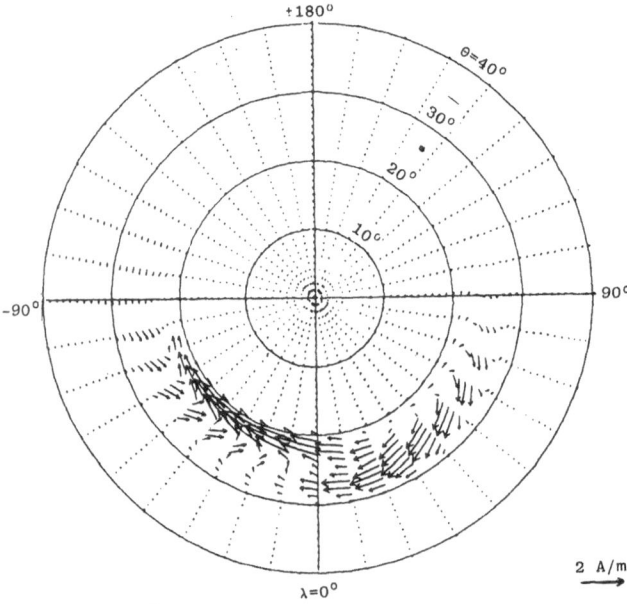

b

Fig. 4.4. a Electric equipotential distribution (3-kV contour interval): b distribution of the ionospheric current vectors (2 A/m vector scale). (Kamide and Matsushita 1979a)

Effects of Double Field-Aligned Currents

Satellite observations revealed the existence of a pair of field-aligned current sheets at all local times even during relatively quiet periods (Zmuda and Armstrong 1974a; Yasuhara et al. 1975; Iijima and Potemra 1976). The observed current directions in the morning sector are downward and upward on the poleward and equatorward sides of the auroral oval, respectively, and are reversed in the evening sector (see Sect. 2.1.6). The field-aligned currents of the poleward half are, in general, more intense than the equatorward half currents. In order to see how the global field and current patterns vary for such a configuration of double field-aligned currents, the upward current in the equatorward half of the morning sector oval and downward current in the equatorward half of the evening oval are added to the previous simple configuration. The peak current density of the field-aligned current of the poleward half is taken to be

$$j_{\parallel 0}^{P} = 2.0 \times 10^{-7} \, \text{A/m}^2.$$

The total current is 3.8×10^5 A. The maximum density of the equatorward half currents is assumed to be

$$j_{\parallel 0}^{E} = 1.0 \times 10^{-7} \, \text{A/m}^2.$$

The total current can therefore be calculated as 2.1×10^5 A. Here, the superscripts P and E stand for poleward half and equatorward half, respectively, of the field-aligned current region.

Figure 4.5a shows the calculated potential pattern. It appears that only the effect of the net field-aligned currents could affect the field pattern in the polar cap. That is since the densities of the poleward half field-aligned currents are assumed to be twice those of the equatorward half currents at all local times, the polar cap potential is influenced only by the field-aligned currents of the poleward half. Perhaps the most striking point is that the potential contours are confined within relatively high latitudes in the present case, indicating that the field-aligned currents of the equatorward half effectively shield the electric field. It is obvious that the relative strength of these two field-aligned current systems determines how rapidly the field is shielded in the vicinity of the equatorward edge of the auroral latitudes (see Sect. 5.3 for details of recent observations). Note that although the relative ratio of the two field-aligned current systems is given in the series of calculations in this section, it is not independent of global magnetospheric dynamics. In particular, as discussed in Section 3.2, the plasma population in the magnetosphere modifies the magnetospheric convection coupled with the ionospheric characteristics. The shielding efficiency of the high-latitude electric field depends in some crucial way on the relative importance between the ionospheric and magnetospheric Hall conductivities (Vasyliunas 1972).

Figures 4.5b shows the vector distribution of the corresponding ionospheric currents. These are characterized by the eastward electrojets in the afternoon sector and the westward electrojets in the midnight-morning sector, where the maximum intensity is almost equal for the two directions. The eastward electrojet has a northward component as well, while the westward electrojet has a southward component except in the midnight hours. The ionospheric current in the

ELECTRIC POTENTIAL

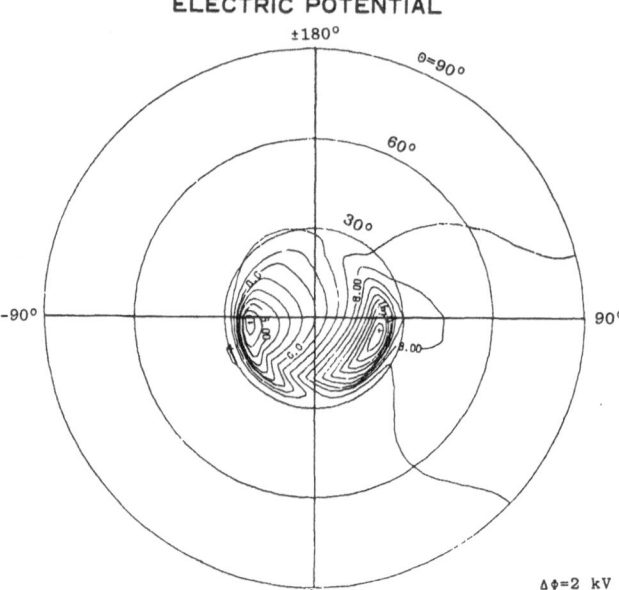

a

IONOSPHERIC CURRENT VECTORS

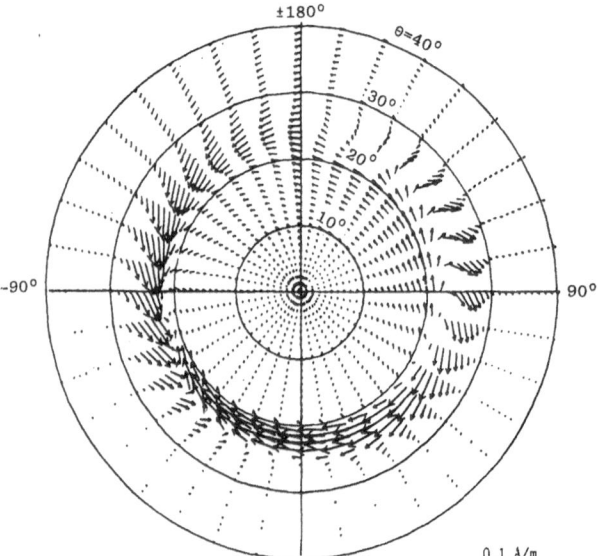

b

Fig. 4.5. a Electric equipotential distribution (2-kV contour interval); **b** vector distribution of the ionospheric current

EQUIVALENT IONOSPHERIC CURRENT VECTORS

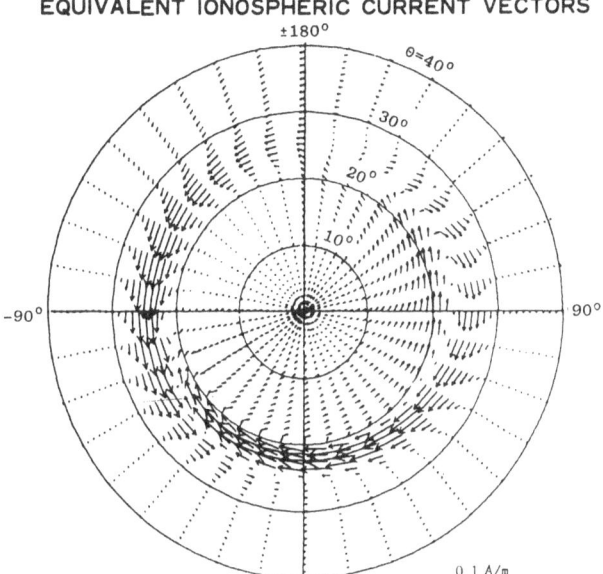

Fig. 4.5. c The equivalent ionospheric currents (0.1 A/m vector scale) for the quiet period with double field-aligned currents at all local times. (Kamide and Matsushita 1979a)

polar cap, especially in the nightside, is very small, less than 0.01 A/m. It is interesting to note that contrary to such a tendency, the equivalent ionospheric currents at the auroral latitudes (not shown here) lie in an almost perfect east-west direction, an indication that the field-aligned currents act to cancel the ground magnetic effects of the north-south component of the ionospheric currents.

It is clear that the non-uniform conductivity distribution makes the entire calculation process laborious, compared to some of the earlier studies along the same line (e.g. Lyatsky et al. 1974). This non-uniform conductivity distribution is found to cause considerable distortions in the global electric field pattern and corresponding current system, in agreement with observations during disturbed periods. Although one tends to consider that the downward (or upward) field-aligned currents coincide with the high (or low) potential point, this is true only when the conductivity is assumed to be uniform. The displacement between the point of the maximum total electric field and the point of the peak field-aligned currents depends on the conductivity gradients as well as the distribution pattern of the field-aligned currents.

4.1.3 Substorm Times

A model calculation for the maximum phase of medium-sized substorms or a "typical" substorm period is presented. Note that actual substorms exhibit highly variable processes in the magnetosphere and ionosphere system. Based on recent satellite observations, the main characteristics of the current flow during sub-storm periods can be summarized as follows: (1) the field-aligned currents are confined to the region of the auroral oval. (2) The field-aligned currents are intensified significantly during substorms. (3) The intensities of the upward and downward currents are generally not equal. To characterize these features, it is assumed that the distribution functions of the field-aligned current density are

$$j_\parallel^P = \pm j_{\parallel 0}^P \exp\left(-\frac{(\theta - \theta_\theta^P)^2}{(D_\theta^P)^2} - \frac{(\lambda \pm \lambda_0^P)^2}{(D_\lambda^P)^2} \right)$$

$$j_\parallel^E = \mp j_{\parallel 0}^E \exp\left(-\frac{(\theta - \theta_0^E)^2}{(D_\theta^E)^2} - \frac{(\lambda \pm \lambda_0^E)^2}{(D_\lambda^E)^2} \right),$$

(4.19)

where P and E stand for the poleward and equatorward portions of the field-aligned currents, respectively, and the upper or lower sign is taken for positive or negative values of λ. For details of the parameters, see Kamide and Matsushita (1970b).

The total downward field-aligned currents $I_\parallel^P$ and $I_\parallel^E$ amount to:

$$I_\parallel^P = 1.9 \times 10^6 \, \text{A} \quad \text{and} \quad I_\parallel^E = 1.1 \times 10^6 \, \text{A}$$

for the field-aligned currents of the poleward and the equatorward half respectively.

Simultaneous observations of large-scale auroras and field-aligned currents by the DMSP and TRIAD satellite investigated, for example, by Kamide and Rostoker (1977) have indicated that there are essentially, at least, four different regions with different auroral luminosities corresponding to different directions and intensities of the field-aligned currents. To simulate such circumstances, four conductive regions are taken into account in the model. As shown in Fig. 4.1, we divide the entire conductive area into four regions. In each of the four regions the height-integrated conductivities are assumed to have Gaussian distribution functions given by

$$\Sigma_{\theta\lambda}^i = \Sigma_{2m}^i \exp\left[-\frac{(\theta - \theta_0')^2}{(D_\theta)^2} - \frac{(\lambda - \lambda_\theta')^2}{(D_\lambda)^2} \right]$$

(4.20)

$$\Sigma_{\theta\theta}^i = \tfrac{1}{2}\Sigma_{\theta\lambda}^i / \sin \chi$$

$$\Sigma_{\lambda\lambda}^i = \tfrac{1}{2}\Sigma_{\theta\lambda}^i \sin \chi,$$

where $i =$ I, II, III, and IV. The maximum conductivity in each region is taken us

$$\Sigma_{2m}^I = 10 \, \text{S} \qquad \Sigma_{2m}^{II} = 40 \, \text{S}$$
$$\Sigma_{2m}^{III} = 20 \, \text{S} \qquad \Sigma_{2m}^{IV} = 20 \, \text{S}$$

Region I represents the poleward half of the morning auroral oval, where the downwardf field-aligned current is usually observed and auroral activity is rather weak. It is region II in which the most active auroras, including the westward traveling surge and torch structure, are usually observed during substorms. Therefore, the largest conductivity value is assumed in this region. In region III an upward field-aligned current is assumed which is confined to the bright auroras there which are, however, not as bright as region II auroras. The diffusive

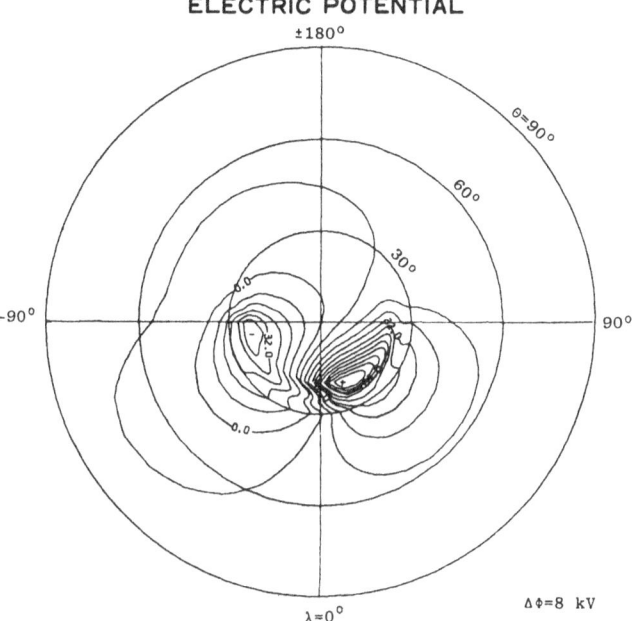

Fig. 4.6. Electric equipotential distribution (8-kV contour interval) over the northern hemisphere ($\theta = 90°$ is the equator and $\lambda = 0°$ is the midnight meridian) for a typical substorm. (Kamide and Matsushita 1979b)

and rather uniform aurora is typically observed in region IV, which is co-located with the downward current.

Figure 4.6 shows the calculated potential distribution. If the potential at the pole is assumed to be zero as one of the boundary conditions, the highest and lowest potential values are 88 and $-40\,\text{kV}$, respectively. The large difference between the magnitudes of these values indicates how strongly the conductivity gradients influence the overall potential distribution. There is also a considerable distortion of the equipotential contours within the nightside conductivity inhomogeneity, which produces an accumulation of the space charges. A comparison of this potential pattern for the typical substorm model with Fig. 4.2 for quiet periods shows how intense auroral enhancement can change the pattern of equipotential contours during very quiet times. For example, the potential centers undergo a significant shift from quiet periods to substorms; the high potential contours move toward midnight, while the low contours move away from midnight.

The distribution of the ionospheric current vectors is shown in Fig. 4.7a, where we note the following important features: (1) the eastward electrojet flows in the equatorward half of the evening auroral belt, while the westward electrojet flows in wider regions in the evening and morning sectors. (2) The main part of these electrojets is supplied by the assumed field-aligned currents, because the ionospheric current in the polar cap and mid-latitudes is very small. (3) The westward electrojet appears to have two peaks, one at premidnight and one in

IONOSPHERIC CURRENT VECTORS

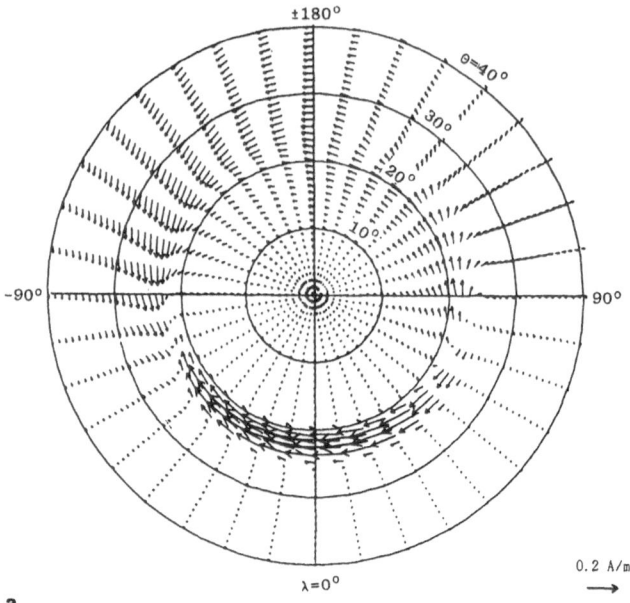

a

EQUIVALENT IONOSPHERIC CURRENT SYSTEM

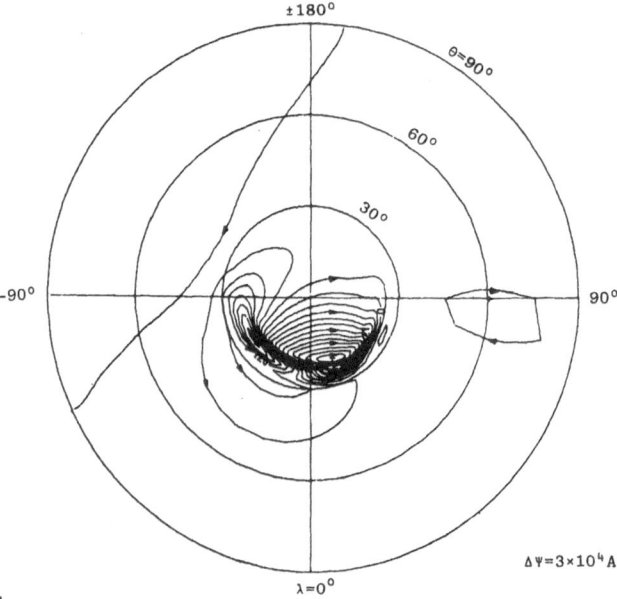

b

Fig. 4.7. a Vector distributions (0.2-A/m vector scale) for a typical substorm of the ionospheric currents at latitudes higher than 50°; **b** the corresponding equivalent inospheric current system in the northern hemisphere (3×10^4 A interval). (Kamide and Matsushita 1979b)

the early morning hours. The maximum current density of the westward electro-jet in the premidnight sector is produced primarily by the assumed high conduc-tivity associated with a bright aurora. On the other hand, the westward electrojet in the morning sector occurs mainly as a result of the large electric field there, although the contribution from the enhanced conductivity is not negligible. The morning electrojet has a considerable southward component, which connects the downward field-aligned current (to the north) and the upward current (to the south) via the Pedersen current. (4) The eastward electrojet is about one-third less intense than the westward electrojet, a result in good agreement with stati-stical results given by Kamide and Fukushima (1972). (5) The eastward electrojet has a northward component, which becomes more intense with the progress of local time; even a totally northward current is seen in the premidnight sector. This means that a significant fraction of the eastward electrojet turns northward and eventually joins the westward electrojet (Baumjohann et al. 1980).

Figure 4.7b presents the distribution of the equivalent current streamlines. There are two main differences between Fig. 4.7a and b. First, the magnitude of the electrojet is less in the equivalent currents than it is in the real ionospheric current, an indication that the field-aligned currents act to cancel part of the iono-spheric current effects. Second, although the electrojets have northward and southward components in the evening and morning sectors, respectively, the equivalent electrojets flow almost in the east-west direction. This indicates that the small or zero declination component perturbation in ground magnetic obser-vations at auroral latitudes during a substorm does not necessarily mean that electrojets are flowing only in the east-west direction. It is also interesting to note that in the polar cap, particularly near the auroral latitudes in the dark sector, the equivalent current vectors are directed from dusk to dawn, consisting of the equivalent "return" currents of the auroral electrojets. Since there is little iono-spheric current flow in the region, these apparent return currents are the effects of the field-aligned currents. Thus, the sunward magnetic perturbations in the polar cap, typically observed during substorms, are indeed produced by the field-aligned currents, not by ionospheric currents. An interesting point here is that even the eastward electrojet in the evening sector looks like the return current from the westward electrojet flowing at higher latitudes of the same meridian.

Kamide and Matsushita (1979b) simulated the variability of the ionospheric electric fields and currents in relation to the distribution of the field-aligned currents, which varies considerably during magnetospheric substorms. Changes in several parameters to a realistic degree of the typical substorm model have been assumed. These changes are effects of intensity variations and location shifts for field-aligned currents; effects of electric conductivity variations and seasonal changes; and effects of additional field-aligned currents and an expanded auroral oval.

4.1.4 Cusp Structure

Magnetospheric convection and the associated electrodynamics processes on the dayside ionosphere are strongly influenced by conditions in the solar wind, in particular by the orientation of the IMF (see Sect. 2.1.3). The dayside cusp separates geomagnetic field line that close on the dayside of the magnetosphere from those that are pulled back into the magnetotail. A set of field-aligned currents that complete the high-latitude ionospheric currents plays a key role in this interesting region of direct contact between the solar wind and the ionosphere. This section deals with simple models to examine how well simulations can reproduce the observed consequence of ionospheric electric fields.

Rich and Kamide (1983) and Banks et al. (1984) treated this problem of the distribution of high-latitude, dayside electric fields and currents in almost identical ways. Assuming that the high-latitude electric field is a consequence of direct

Ionospheric Electric Potential Pattern

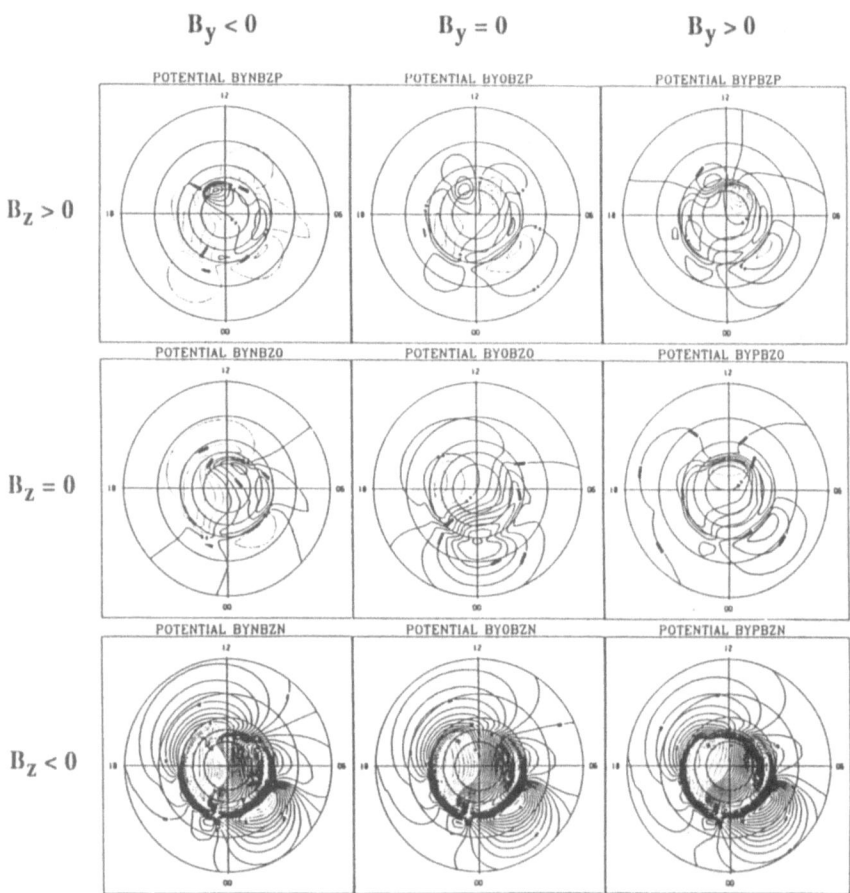

Fig. 4.8. a Polar plot of the distribution of field-aligned currents used in the computer model of Clauer and Friis-Christensen (1988). The *shaded region with solid contours* indicates currents into the ionosphere and the *unshaded region with dashed contours* indicates currents out of the ionosphere in the northern hemisphere. The contour interval is at 0.5 μ A/m²

coupling between the interplanetary electric field and the dayside ionosphere through field-aligned currents, their simulation codes solve the steady-state equation for current conservation by using models of field-aligned currents and the height-integrated ionospheric conductivities, including seasonal, geomagnetic activity, and solar wind changes. The IMF influence can be described by a decomposition of the dayside field-aligned currents into systems which are separately controlled by the IMF B_y and B_z components.

Ionospheric Distribution of Field-Aligned Current

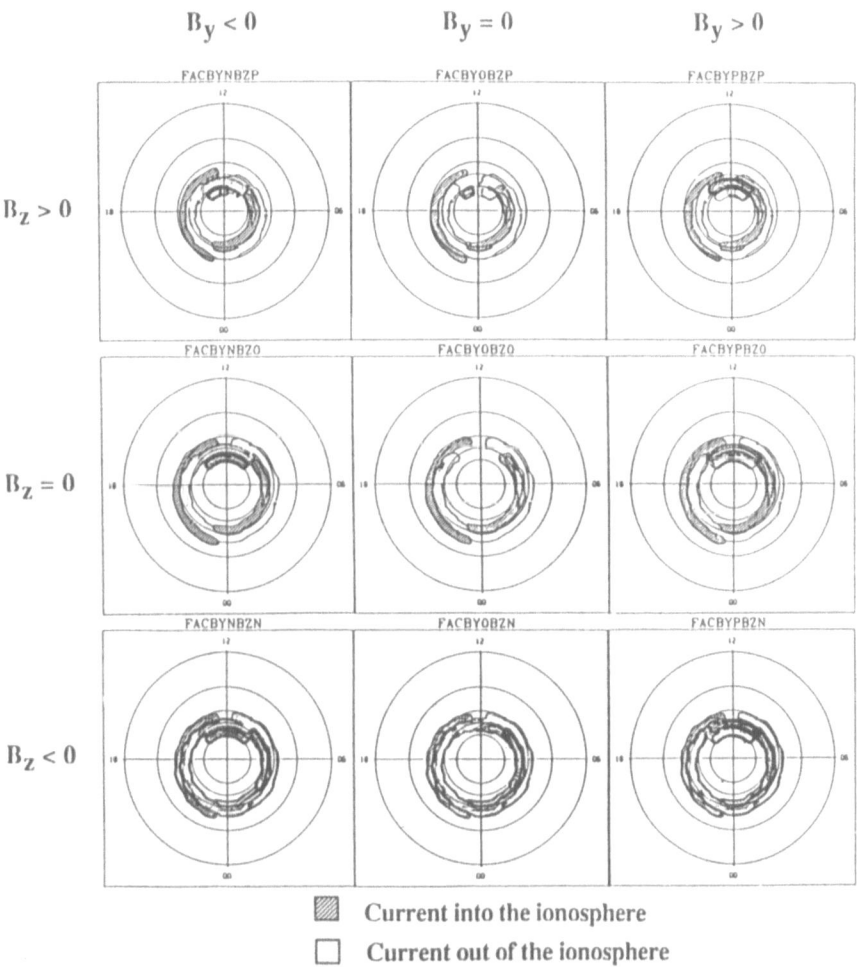

Fig. 4.8. b Polar plot of the ionospheric potential distribution associated with each of the current distributions shown in a. The contour interval is 8 kV. The zero potential curve crosses the center of the polar cap; *solid contours* indicate positive values, and *dashed contours* indicate negative values. (Clauer and Friis-Christensen 1988)

In a more recent study, Clauer and Friis-Christensen (1988) refined the modeling scheme of Banks et al. (1984) and Clauer and Banks (1986). The refinement addresses more carefully the situation for IMF both northward and southward, by including latitudinal changes in the field-aligned current systems when IMF B_y changes the sign. The nine plots in Fig. 4.8a show the variability with respect to the IMF of the field-aligned current distributions used in Clauer and Friis-Christensen (1988). In their model, it was assumed that there is a set of independent dayside field-aligned current systems, each separately controlled by the IMF B_y and B_z components. The DPY system (Friis-Christensen and Wilhjelm, 1975; Wilhjelm et al. 1978) consists of a pair of oppositely directed current sheets oriented longitudinally across the noon meridian (see Sect. 2.1.3). Associated with changes in the IMF B_y component, the DPY system is derived. One of the current sheets is co-located with the Region 1 current, and the other is located poleward of it. For southward IMF ($B_z < 0$), on the other hand, the field-aligned current system, called DPZ, is a current wedge with current flowing from the prenoon magnetopause into the ionosphere at the latitude of the Region 1 current and out of the ionosphere in the postnoon Region 1 current. During northward IMF, the field-aligned DPZ system (i.e. the NBZ system in Sect. 2.1.3) is a similar current wedge with opposite current directions, but located near the poleward boundary of the cusp. The pattern in the center in Fig. 4.8a shows the so-called baseline ($B_z = B_y = 0$) configuration, in which no IMF effect is involved.

Figure 4.8b shows the patterns of the resultant electric potential corresponding to these nine distributions of the field-aligned current systems. In the simulation, the dayside and nightside Region 1 and Region 2 currents are assumed to distribute continuously along the auroral oval, although these currents may in fact be separated, being produced by separate sources, as indicated by Friis-Christensen et al. (1985). This diagram is quite useful for a comparison with a similar simulation result of Friis-Christensen et al. (1985), which was derived from a statistical data set of ground magnetometer data sorted for different IMF conditions and from the magnetogram-inversion technique (see Sect. 4.2). The most noteworthy point, however, is that comparing Fig. 4.8b with dayside radar observations, the study of Clauer and Friis-Christensen (1988) presented a rapid, almost immediate, response of the high-latitude electric field to north-south changes in the IMF. This character is somewhat inconsistent with the observations of Wygant et al. (1983) in that it requires several hours for the polar cap potential (i.e. for the "turning-off" of the reconnection process) to respond to northward IMF.

4.2 Magnetogram-Inversion Technique

Unveiling the dynamics of the configuration of the three-dimensional current systems is of great importance in understanding the degree to which the magnetosphere and the polar ionosphere are electrically coupled. It is known that in reality, ionospheric currents, such as the intense auroral electrojets, are connected to field-aligned currents in a very complicated fashion, reflecting the variability of the degree of the magnetosphere–ionosphere coupling and causing a variety of different types and patterns of auroras and ground magnetic perturbations. As pointed out by Potemra (1987), however, the debate between Chapman and Birkeland, concerning the existence of the field-aligned currents, continued until the mid-1960s (Zmuda et al. 1966), primarily because it is not possible to determine uniquely the separate effects of ionospheric and field-aligned currents only from ground magnetometer observations, on which the past studies have heavily relied.

Historically, ground-based magnetic records have been widely used to examine physical processes occurring in the magnetosphere and the ionosphere, since the magnetic data obtained at the Earth's surface include valuable information about a variety of source currents flowing in the near-Earth environment, including the ionosphere and magnetosphere, the magnetopause, and the Earth's interior (see Nishida 1978). However, for that very reason, namely, that the data contain too much information, it has been a serious problem to evaluate the relative importance of these currents in generating the particular patterns of global or local magnetic perturbations that we are attempting to study. One of the important problems in geomagnetism has been to ascertain the relative importance of these different source currents, since it is in principle impossible to determine uniquely the three-dimensional distribution of the current system solely from magnetic observations made at the Earth's surface (Chapman 1935).

The International Magnetospheric Study (IMS, 1976–1979), during which systematic data from ground-based magnetometer arrays were collected (Lanzerotti et al. 1976), coincided with the development of several computational techniques called the magnetogram-inversion methods; see reviews, for example, by Rees (1982), Kamide (1982, 1988), Feldstein and Levitin (1986), Glaßmeier (1987), and Mishin (1990). These numerical schemes are designed to compute not only the global distribution of both ionospheric and field-aligned currents, but also of ionospheric electric fields and the Joule heating rate (Fayermark 1977; Kisabeth 1979; Mishin et al., 1979; Baumjohann et al. 1981; Kamide et al. 1981; Glaßmeier 1987). Once we understand properly the origins of ground magnetic perturbations in terms of various source currents, ground-based observations have an advantage over "more direct" measurements by radar and satellite, since variations in the geomagnetic field are being monitored continuously at a relatively large number of fixed points on the Earth's surface. This contrasts to the intrinsic ambiguity in determining the global three-dimensional current system on an individual basis from single satellite passes, which have difficulties separating temporal and spatial changes.

4.2.1 Essence of the Scheme

The purpose of this section is to describe the essence of the magnetogram-inversion scheme, along with its assumptions and limitations. Although several magnetogram-inversion techniques start with the same set of basic equations, the practical procedures as well as the points of emphasis vary considerably, depending on different algorithms. In the so-called Forward modeling to infer the global distribution of three-dimensional current flow (Kisabeth 1979), the high-latitude ionosphere is divided into a large number of cells, each of which is associated with two elementary field-aligned current systems: east-west closing and north-south closing currents across the cell in the ionosphere. The computer code in this modeling is designed to find a set of these two currents for all cells that best reproduces the input data of the ground-based magnetic perturbations.

In a sense, the Forward method is a trial-and-error procedure. Although the Forward method does not take the ionospheric conductivities explicitly as input, the University of Münster group (e.g. Baumjohann et al. 1981) uses an assumed conductivity distribution, representing particular auroral forms, along with observations of the ionospheric electric field in a localized region. The conductivity is then changed until sufficient agreement is obtained between the calculated and observed ground magnetic perturbations.

The other major algorithms (e.g. Fayermark 1977; Mishin et al. 1979; Kamide et al. 1981) for obtaining the three-dimensional current system and ionospheric electric fields require, first of all, that an ionospheric equivalent current system can be adequately derived from an array of ground-based magnetic field measurements. The equivalent current system is a toroidal horizontal sheet current $\mathbf{J}_T$ assumed to be flowing in a shell at a 110-km altitude, whose associated magnetic field matches the external portion of the observed magnetic variation field $\mathbf{b}$. The toroidal current can be expressed in terms of an equivalent current function ψ: see Eq. (4.12). In the lower atmosphere, where negligible electric current flows, the magnetic variation can be expressed in terms of a magnetic potential V as

$$\mathbf{b} = - \text{grad } V. \tag{4.21}$$

The external portion of V is uniquely related to ψ by straightforward mathematical relations (see Chapman and Bartels 1940, p. 631). However, the practical calculation to separate the magnetic potential into external and internal origins and to extend the external portion toward the ionospheric sources is quite complicated, reflecting primarily the inevitable consequence of data noise and the limited station coverage (see Mersmann et al. 1979).

With a given equivalent current function ψ and given conductivities, one is able to derive the ionospheric electric field, horizontal current, and field-aligned current under the following simplifying assumptions: (1) the electric field is electrostatic. (2) Geomagnetic field lines are effective equipotentials, i.e. there is no "parallel" electric field. (3) The dynamo effects of ionospheric winds can be neglected. (4) The magnetic contributions of magnetospheric ring currents, magnetopause currents, and magnetotail currents to the equivalent current function can be neglected. (5) Geomagnetic field lines are effectively radial.

As described in Eq. (4.8), the height-integrated horizontal ionospheric current $\mathbf{J}$ can be expressed as the sum of the toroidal (equivalent) current and a "potential" current $\mathbf{J}_T$ as

$$\mathbf{J} = \mathbf{J}_T + \mathbf{J}_P, \tag{4.22}$$

where the second component can be written in terms of a current potential τ. This is a special case of the well-known Helmholtz theorem for vector analysis. The potential component can be considered as a closing current for field-aligned currents. The requirement that the three-dimensional current be divergence-free means that the field-aligned current density $j_\parallel$ (positive downward) satisfies the relation

$$j_\parallel = \text{div }\mathbf{J} = \text{div }\mathbf{J}_P, \tag{4.23}$$

since $\mathbf{J}_T$ is by definition divergence-free. The current system represented by $j_\parallel$ and $\mathbf{J}_P$ together produces no ground magnetic variation under the assumption that the toroidal component of $\mathbf{J}$ is just the equivalent current system. That is, the equivalent current system estimated from ground magnetic perturbations produced by $\mathbf{J}$ and $j_\parallel$ (in other words, $\mathbf{J}$ and $-\mathbf{J}_P$) is identical to $\mathbf{J}_T$ (see discussion in Sect. 4.1).

Since the horizontal ionospheric current is related to the electric field $\mathbf{E}$ by

$$\mathbf{J} = (\Sigma)\mathbf{E} = \Sigma_P \mathbf{E} + \Sigma_H \mathbf{E} \times \mathbf{n}_r, \tag{4.24}$$

where Σ_P and Σ_H are the height-integrated Pedersen and Hall conductivities, Eq. (4.22) is then rewritten as

$$-(\Sigma)\text{grad }\Phi = -\text{grad }\tau - \text{grad }\psi \times \mathbf{n}_r, \tag{4.25}$$

where $\mathbf{J}_P = -\text{grad }\tau$. A partial differential equation for the electrostatic potential Φ in terms of ψ can be obtained by taking the curl of Eq. (4.25) as

$$\text{curl}(\text{grad }\psi \times \mathbf{n}_r) = \text{curl}((\Sigma)\cdot\text{grad }\Phi). \tag{4.26}$$

The magnetogram-inversion technique requires knowledge of the height-integrated Pederson and Hall conductivities. Note, in particular, that the KRM numerical scheme (Kamide et al. 1981) accepts essentially any anisotropic distribution of the conductivities. In spherical coordinates θ (colatitude) and λ (east longitude), one obtains

$$A\frac{\partial^2\Phi}{\partial\theta^2} + B\frac{\partial\Phi}{\partial\theta} + C\frac{\partial^2\Phi}{\partial\lambda^2} + D\frac{\partial\Phi}{\partial\lambda} = F, \tag{4.27}$$

where the coefficients of this second-order differential equation are given by

$$A = \sin\theta\cdot\Sigma_H$$

$$B = \frac{\partial}{\partial\theta}(\sin\theta\,\Sigma_H) + \frac{\partial}{\partial\lambda}\Sigma_P$$

$$C = \Sigma_H/\sin\theta$$

$$D = \frac{\partial}{\partial\theta}\Sigma_P - \frac{\partial}{\partial\lambda}\left(\frac{\Sigma_H}{\sin\theta}\right)$$

$$F = \frac{\partial}{\partial\theta}\sin\theta\frac{\partial\psi}{\partial\theta} + \frac{1}{\sin\theta}\frac{\partial^2\psi}{\partial\lambda^2}.$$

With a given current function ψ (or F) and given conductivities and their gradients (in $A, B, C,$ and D), the electric potential Φ can be solved for certain boundary conditions. It is important to point out that if all the conductivity gradients are neglected, Eq. (4.27) is reduced to the simple Poisson equation for

Φ as

$$\frac{\Sigma_H}{\sin\theta}\frac{\partial}{\partial\theta}\left(\sin\theta\frac{\partial\Phi}{\partial\theta}\right)+\frac{\Sigma_H}{\sin^2\theta}\frac{\partial^2\Phi}{\partial\lambda^2}=\frac{1}{\sin\theta}\frac{\partial}{\partial\theta}\left(\sin\theta\frac{\partial\psi}{\partial\theta}\right)+\frac{1}{\sin^2\theta}\frac{\partial^2\psi}{\partial\lambda^2}, \tag{4.28}$$

indicating that the solution can be given by

$$\Phi=\psi/\Sigma_H. \tag{4.29}$$

This simply means that the world pattern of the electric potential is exactly the same as that of the equivalent ionospheric currents which are derived from ground magnetic perturbations, and that only the constant Hall conductivity is important in estimating the electric potential from the equivalent current system. This relationship has long been employed in many aspects of geomagnetism (see, for example, Kern 1966; Gurevich et al. 1976; Levitin et al. 1982).

Once the electric potential is obtained by solving Eq. (4.27) numerically, the components of the height-integrated ionospheric current can be readily calculated by Eq. (4.24). By inserting the ionospheric currents into Eq. (4.23), it is possible to derive the distribution of the field-aligned current density, obtaining the three-dimensional current system.

Note that most of the algorithms of the magnetogram-inversion techniques assume that geomagnetic field lines are perpendicular to a "flat" ionosphere, which simplifies drastically the mathematical development in solving the equations involved. However, such assumptions may limit their application only to the highest-latitude regions (see also Tamao 1986). Sun et al. (1985) attempted to relax this assumption and developed an iteration scheme, starting with the algorithm put forward by Kamide et al. (1981), to calculate the three-dimensional current system in which the field-aligned currents flow along realistic dipolar field lines. Thus, the iteration method described in Sun et al. (1985) can be considered an extension of the KRM code as well as a test of the accuracy of the KRM code in its original version.

The main results obtained from the new scheme of Sun et al. (1985) are listed as follows: (1) the iteration scheme described here provides an improvement of the KRM algorithm by assuming dipole field lines for the field-aligned currents. The "curved" field-aligned current system contributes approximately 5–15% of the observed total magnetic field at auroral latitudes. (2) The correction to the ionospheric currents and field-aligned currents based on the scheme is found to be 5–20% around the auroral zone. (3) The correction to magnetic field values in the mid-latitude region ($< 60°$) is 20–30%, although the absolute error values in nT are smaller there.

On the other hand, in attempting to make use of information on the ionospheric electric field in relatively localized regions, such as northern Scandinavia, the Untiedt algorithm (Untiedt et al. 1990) takes a somewhat different approach. In a Cartesian coordinate system with the x-, y-, and z-axes pointing to the north, the east, and downward, respectively, Eq. (4.27) is identical to

$$\Sigma_H\left(\frac{\partial^2\Phi}{\partial x^2}+\frac{\partial^2\Phi}{\partial y^2}\right)+\left(\frac{\partial\Sigma_H}{\partial x}-\frac{\partial\Sigma_P}{\partial y}\right)\frac{\partial\Phi}{\partial x}+\left(\frac{\partial\Sigma_P}{\partial x}+\frac{\partial\Sigma_H}{\partial y}\right)\frac{\partial\Phi}{\partial y}=f(x,y), \tag{4.30}$$

where the values of $f(x, y)$, equivalent to F in Eq. (4.27), can be calculated directly from ground magnetometer data. The essence of the Untiedt et al. method is that it obtains the Hall and Pedersen conductivities instead of the electrostatic potential Φ, since Σ_H and Σ_P can be solved in Eq. (4.30) through a first-order (not second-order) differential equation. Using the electric field and the Hall conductivity, Eq. (4.30) can be rewritten as

$$P\frac{\partial\Sigma_H}{\partial x} + Q\frac{\partial\Sigma_H}{\partial y} + R\Sigma_H = f(x, y), \tag{4.31}$$

where

$$P = \mathbf{E}_x + \frac{\mathbf{E}_y}{\alpha} \qquad Q = \mathbf{E}_Y - \frac{\mathbf{E}_x}{\alpha} \qquad R = \frac{\partial P}{\partial x} + \frac{\partial Q}{\partial y},$$

provided that a conductivity ratio $\alpha = \Sigma_H/\Sigma_P$ is specified. The Untiedt et al. algorithm has been applied to the Harang discontinuity, for which data from the Scandinavian magnetometer array and simultaneous electric field observations from STARE radar were available (Segatz 1985; Untiedt et al. 1990).

4.2.2 Advantages and Limitations

Many recent studies have proven that the magnetogram-inversion technique is a powerful tool for quantitatively evaluating the three-dimensional current systems responsible for magnetic disturbances that occur at high latitudes associated with auroral displays (Akasofu and Kamide 1985). These "remote-sensing" schemes have also been shown to be quite useful in discussing magnetosphere–ionosphere coupling and in providing basic information, such as the electrostatic potential at the polar cap boundary, numerical modeling of magnetospheric plasma processes (Wolf and Kamide 1983) and thermospheric wind patterns through the Joule heat dissipation from the ionospheric current (Rees 1983). Unlike satellite measurements, the determination of electrodynamic parameters in the high-latitude ionosphere, using the magnetogram-inversion scheme, is not based on in situ data, but rather primarily on indirect magnetic measurements on the Earth's surface. However, with the magnetogram-inversion scheme it is possible to estimate the ionospheric quantities on a global scale (instead of only along satellite orbits) with a time resolution of about 5 min or even less over extended periods of time.

The development of this particular method in the field benefited from considerable progress in other areas, such as in situ electric field observations by satellites, barium releases from sounding rockets, and measurements from balloons and incoherent scatter radar. It would be useful to modify the numerical algorithm in such a way that simultaneous, "more direct" measurements of electric fields, conductivities, and field-aligned currents by satellite and radar could be incorporated into the scheme. Some of the recent progress along this line is accounted for in Section 4.2.4.

Although not a limitation of the method itself, it must be noted that a model of the ionospheric conductivities must be given in the magnetogram-inversion procedure, although there is at present no simple way in which time changes in the conductivities over the entire polar region can be accurately monitored. Several empirical models of the ionospheric conductances have been proposed, for example, by Wallis and Budzinski (1981), Spiro et al. (1982), Robinson et al. (1987), and Fuller-Rowell and Evans (1987). These average models are derived from radar measurements of the electron density and satellite measurements of precipitating electrons. In view of the somewhat high degree to which the estimated electric field depends on the choice of the conductivity distribution, critical comparisons must be made between the results from magnetogram-inversion techniques and those obtained by means of more direct techniques.

4.2.3 Global Distribution of Ionospheric Parameters

The KRM algorithm (Kamide et al. 1981) has been applied to a number of different sets of ground-based magnetometer data for both quiet and disturbed periods (e.g. Kamide et al. 1982b). In this section we demonstrate one case study (Kamide and Baumjohann 1985), where instantaneous, unsmoothed magnetic data from some hundred variably spaced stations are used as input, making it

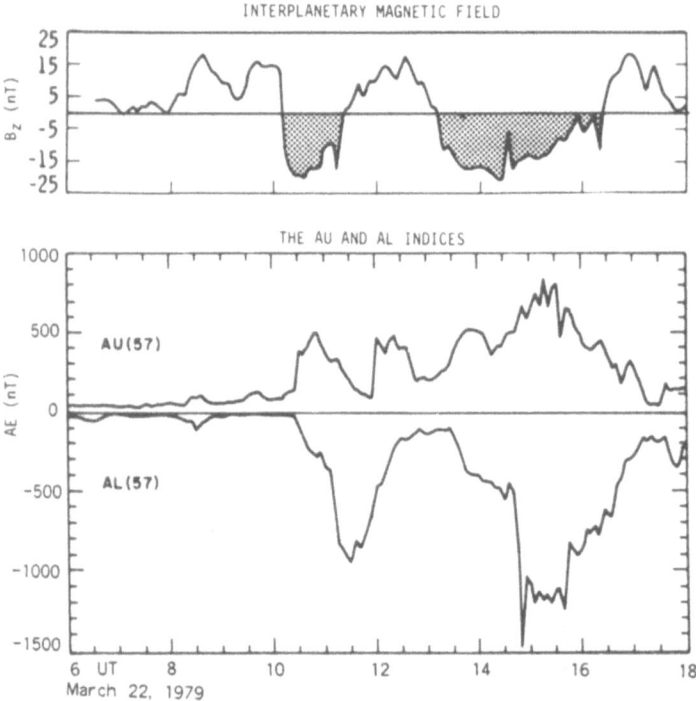

Fig. 4.9. The IMF B_z component measured by IMP 8 and the AU/AL indices for March 22, 1979. The B_z component is in GSM coordinates. (Kamide and Baumjohann 1985)

possible to obtain a "snapshot" of the current system as well as potential patterns over the entire polar region.

The auroral conductance has both solar UV and precipitating particle contributions. For the auroral enhancement, an empirical model based on the work of Spiro et al. (1982) with updated improvements has been used. In this model, the distribution of auroral-induced conductance values depends upon the value of the AE index. The time period studied during CDAW-6 (Coordinated Data Analysis Workshop, Series 6) consists of two intervals of substorm activity, which were selected by the scientific community on the basis of the availability of extensive data sets (McPherron and Manka 1985). Figure 4.9 shows the auroral electrojet indices, AU and AL, for one of the two intervals. The corresponding B_z component of the interplanetary magnetic field (IMF) is also shown. During this interval, two large, relatively isolated substorms occurred. It is evident in Fig. 4.9 that the first substorm is preceded by a lengthy interval of quiet in terms of auroral electrojet activity. The second substorm started just after the recovery of the first substorm to a quiet level.

Associated with the IMF southward turning, AU and AL began to grow at 1020–1022 UT, and reached a peak at 1050 UT. The maximum magnitude of AU at this time was slightly larger than that of AL. This interval was followed by a sudden decrease in AU and an increase in AL intensity at 1055 UT, which can be identified as the expansion onset of the first substorm. The peak of the westward electrojet intensity was recorded at about 1130 UT, and the end of this first substorm occurred at about 1250–1300 UT, which marks an almost full recovery of the geomagnetic activity.

The second substorm followed again a southward IMF turning at 1310 UT, which appeared to enhance electrojet activity starting at 1320–1325 UT. The major expansion of the second substorm was observed at 1436 UT. The AL index indicates that a sharp peak at 1450 UT is followed by approximately 45 min of relatively stable and intense electrojet activity. The recovery of this second substorm started between 1600 and 1700 UT.

As noted recently by Kamide and Akasofu (1983), many important features of detailed structures of the substorm development cannot be fully explained by the AE indices alone. It is difficult to quantify the degree of auroral electrojet activity without having at least several parameters, such as the maximum current density, latitudinal width and the longitudinal extent of the auroral electrojets. Subject to these limitations, in the following we attempt to present the characteristics of the global patterns of electric fields and currents during the major CDAW-6 substorms.

Equivalent Currents

To illustrate complicated (locally) but systematic (on a global scale) structures of substorm development, Fig. 4.10 presents a series of the equivalent current systems from 1030 to 1220 UT. The following three distinct large-scale patterns characterizing different states of the magnetosphere and ionosphere are noted:

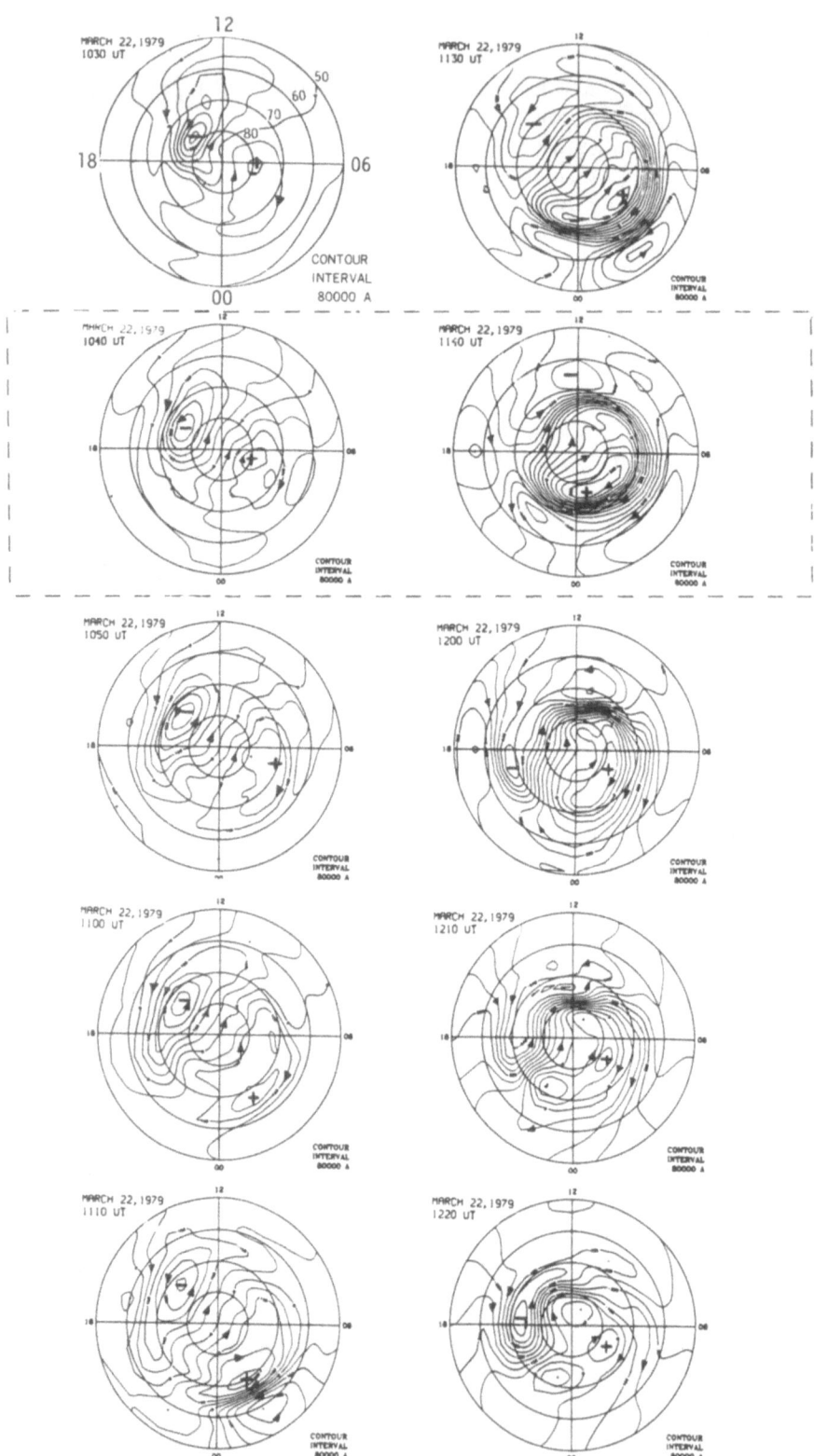

1. Following the southward turning of the IMF, a very clear two-cell current system, designated customarily as the DP 2 system (e.g. Nishida 1968a,b), starts to develop at high latitudes. The current system consists of a clockwise equivalent current in the morning sector and a counterclockwise current in the evening sector, representing presumably the two-cell plasma convection system in the magnetosphere.

2. At 1055 UT, the early morning portion of the clockwise current vortex was suddenly intensified and deformed, signaling the sudden development of an intense substorm westward electrojet. With the enhancement of the intensity of the westward electrojet, its longitudinal extent increased in both eastward and westward directions. As a result of the processes, a one-cell current system prevailed at the maximum epoch of this intense substorm. One may call this system the DP 1 current system. The characteristics of the DP 1 and DP 2 systems before and after the onset of substorm expansion are examined in detail by Clauer and Kamide (1985).

3. During the recovery phase of the substorm, starting at about 1140 UT, a small counterclockwise cell appeared in the premid-night sector and increased its intensity. As a whole, it appears that the DP 1 and DP 2 current systems coexisted. The relative ratio of the two current strengths changes drastically with time; compare the three patterns at 1200, 1210, and 1220 UT for the progressive change in the relative importance of the two basic patterns. In particular, the intensities of the DP 1 and DP 2 currents are nearly equal at 1210 UT, but the global current pattern at 1220 UT is very similar to that before the onset of expansion described in pattern (1). The only notable difference from DP 2 is that in the 1220 UT pattern the nightside portion still seems considerably deformed, perhaps indicating a remnant of the substorm westward electrojet.

The examples shown in this section demonstrate clearly the existence of the two basic modes of the equivalent current systems. The corresponding large-scale patterns in the electric potential and ionospheric and field-aligned currents are also discernible, implying that different physical processes are operating during DP 1 and DP 2 periods. It is important to point out that these two types of current systems can coexist during most substorm intervals and their relative strength varies from time to time (e.g. Baumjohann 1983). This may be why individual current systems seem very complicated.

Potential Pattern

Figure 4.11 shows the calculated potential distributions for 1040 and 1140 UT, which represent the time preceding the major onset of expansion and the time of the maximum epoch of the first substorm, respectively. In the equivalent current signatures, these two epochs have been identified as typical DP 2 and DP 1 current intervals (see Fig. 4.10). An examination of the electric potential pattern for DP 2 intervals is extremely important because the DP 2 system is

Fig. 4.10. Progressive change in the equivalent ionospheric current system (contour interval: 80 000 A) for the first major substorm on March 22, 1979. (After Kamide and Baumjohann 1985)

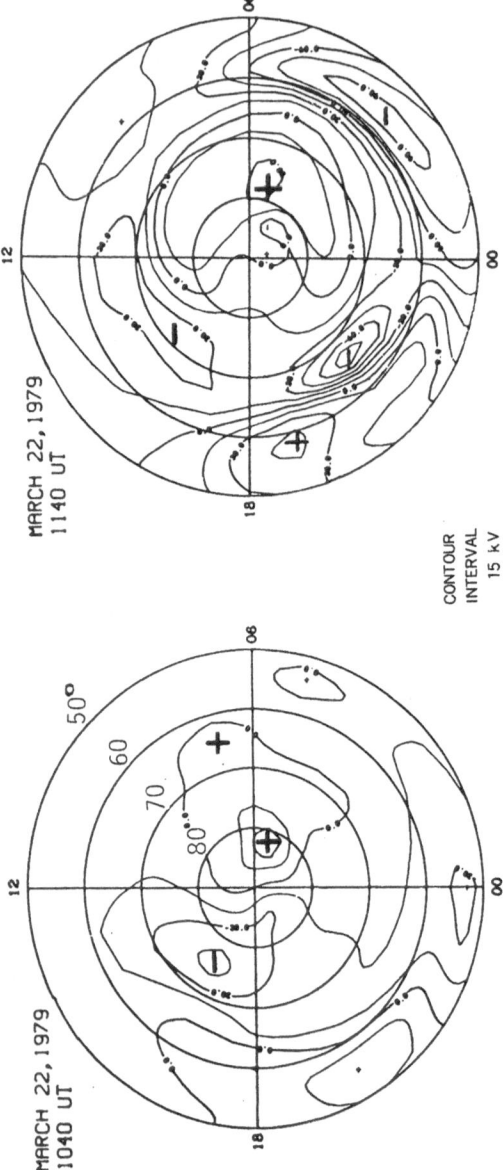

Fig. 4.11. Electric potential (contour interval: 15 kV) for two instances on March 22, 1979. *Left*, The time preceding the expansion onset of the first major substorm; *right*, the maximum epoch of the substorm

believed to result from the electric field associated with large-scale magnetospheric convection. However, since during typical DP 2 periods, substorm activity or ionospheric conductivity enhancements in the midnight sector should be absent, the equivalent current system is quite similar to the potential pattern. Although in the equivalent current system at 1040 UT the demarcation line between the morning and evening vortices is along the 1000–2200 MLT meridian, the line separating the two potential vortices appears to align nearly with the noon-midnight meridian.

The potential distribution during the maximum phase of the substorm at 1140 UT features a two-cell pattern at high latitudes, but the high-potential vortex generating the southward electric field in the morning sector is much larger than its counterpart in the afternoon sector. Note that since in the numerical scheme all field lines are assumed to be vertical to the horizontal ionosphere, the potential values at subauroral latitudes become unrealistically large.

Ionospheric Currents

One of the advantages of the magnetogram-inversion technique is that it is possible to calculate the distribution of ionospheric currents, not equivalent currents, through the assumption of the ionospheric conductance. It has been found by using a number of examples, both statistical and individual cases (e.g. Kamide and Richmond 1982), that the ionospheric current distribution obtained through the inversion scheme is not sensitive to the choice of the conductivity model, unless auroral enhancements are misplaced in the model.

Figure 4.12 shows the distribution of the calculated ionospheric current vectors. Comparing the equivalent and "true" ionospheric currents, one can note that the gross patterns of the two currents are very similar for all the epochs, indicating that the ground magnetic perturbations at auroral latitudes are produced mainly by the ionospheric currents. A significant difference can be found in the current direction in the dayside polar cap. This difference indicates the importance of field-aligned current effects at these local time sectors. It is of great interest to compare the above instantaneous distributions of the ionospheric current system with the average pattern which was determined earlier using the same algorithm but different, averaged data, namely the data set obtained from the Alaska meridian chain (e.g. Akasofu et al. 1980, 1981). The gross features are quite similar between the average and instantaneous distributions. A significant difference appears in the evening sector when the westward traveling surge intensified and the westward electrojet penetrated into the evening sector, a feature indicating that the Harang discontinuity region is subject to very dynamic behaviour during individual substorms.

In the present scheme, the ionospheric current density is computed for every 1° of latitude and 1 h of MLT. By integrating the east-west current density, the total eastward and westward ionospheric currents are deduced at each MLT sector. Figure 4.13 shows the maximum eastward and westward currents throughout the CDAW-6 interval. Plotted along with the total electrojet currents are the *AU* and *AL* indices, which are transferred from Fig. 4.10. From their definition,

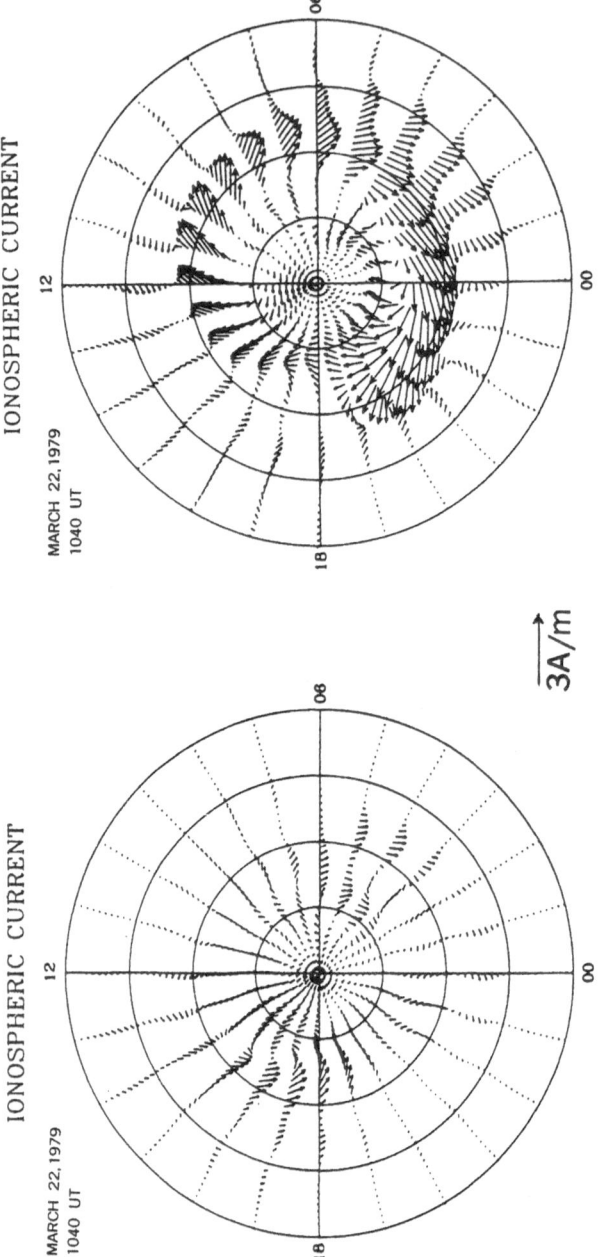

Fig. 4.12. Ionospheric current vectors (3 A/m vector scale) for two instances on March 22, 1979, representing the pre-expansion (*left*) and the maximum epoch (*right*) of the first major substorm

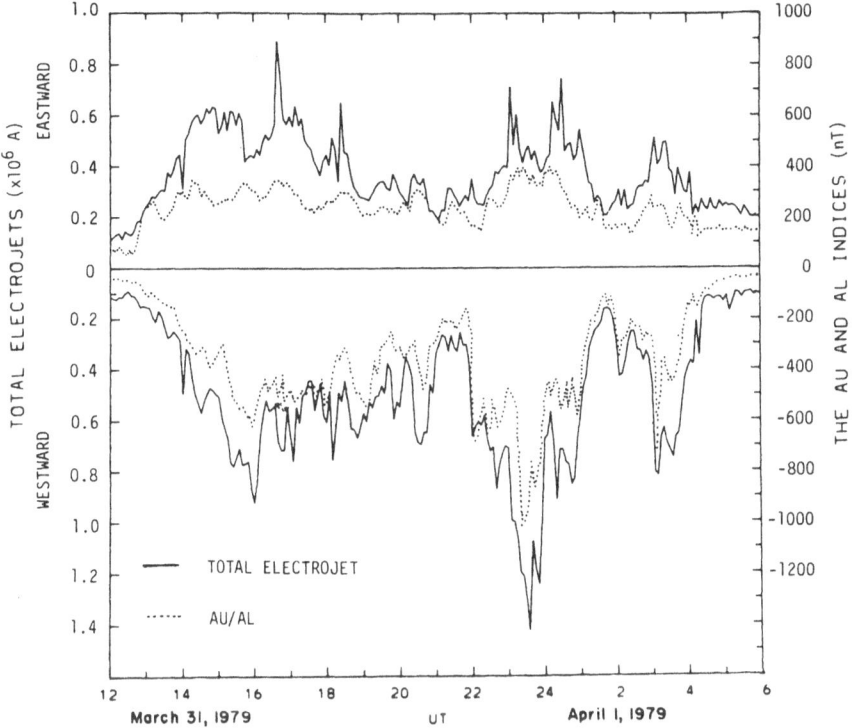

Fig. 4.13. Comparison between the total electrojet current (for eastward and westward, separately) and the AU/AL indices for March 31 and April 1, 1979. (Kamide and Baumjohann 1985)

the AU and AL indices should express the maximum current density of the eastward and westward electrojets, respectively.

The general trends between the AU and the total eastward current, and between AL and the westward current, are similar. The correlation coefficients range from 0.75 to 0.94. The average normalization factors in the vertical scales in Fig. 4.13 are: 500 nT in AU corresponds to 0.75×10^6 A in the total eastward current, whereas 500 nT in AL corresponds to 0.90×10^6 A in the total westward current. This indicates that the average latitudinal width of the westward electrojet is somewhat larger than that of the east-ward electrojet.

Field-Aligned Current

The distribution of field-aligned currents, which are calculated from the divergence of the ionospheric current vectors, appears to be very complicated. Figure 4.14 shows temporal changes in the calculated upward and downward currents that include many local structures. These diagrams represent the maximum phase of the first substorm. The central location and the intensity of the field-aligned currents change significantly during the 10-min interval. It is also impor-

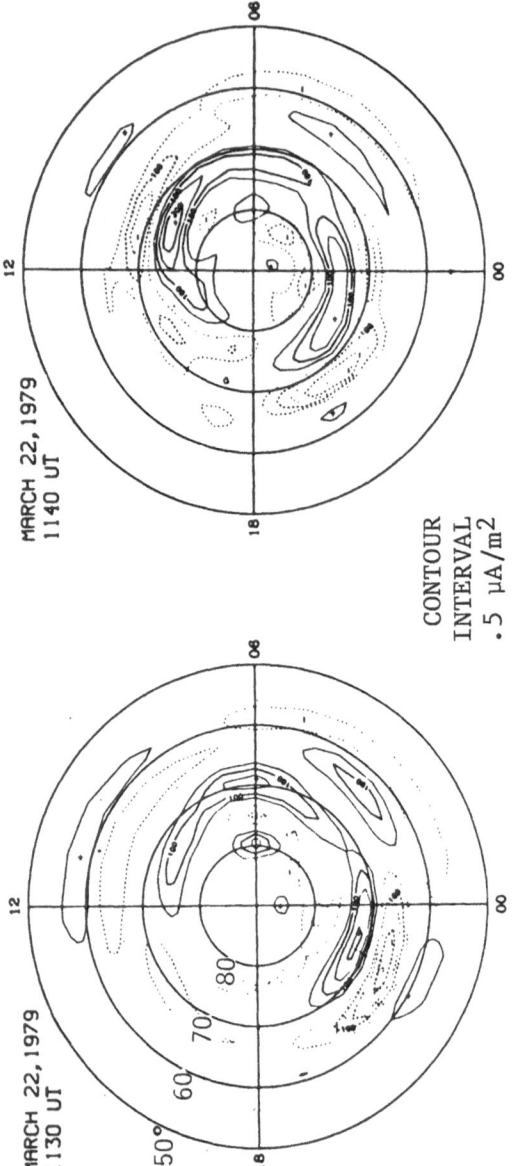

Fig. 4.14. Field-aligned current distribution (contour interval: 0.5×10^{-6} A/m²) for two adjacent times during the maximum epoch of the first major substorm on March 22, 1979

tant to point out that there is great variability in the latitudinal and longitudinal distribution of the field-aligned currents during individual substorms in comparison with the statistical pattern obtained by averaging a large number of satellite measurements (e.g. Iijima and Potemra 1976). In particular, the Region 1 current in the morning sector appears to be split latitudinally into two parts and extends longitudinally into the premidnight hours. The Region 2 current seems to be weak during this interval.

An inspection of a number of the field-aligned current distributions for the entire CDAW-6 period indicates that it is very often difficult without preconception to identify properly even the Region 1 and Region 2 currents. Substorm variations of the current distribution deviate considerably from a simple state in which the large-scale pattern is unchanged and only the intensity changes with the growth and decay of substorm activity. Particularly, during the expansive phase, many local structures of upward and downward field-aligned currents appear and disappear in the dark sector. This may not be too surprising in a sense, because the distribution of the field-aligned currents represents, mathematically, the first derivative of the ionospheric current and the second derivative of the electric potential and, physically, local enhancements of the auroral electrojet currents. It must be noted, however, that the derived field-aligned currents do not reflect the small-scale enhancements of the ionospheric conductivity since only the statistical conductivity model has been employed.

4.2.4 Recent Improvements

It must be noted once again that, unlike rocket and satellite measurements, the derivation of field-aligned currents and other ionospheric parameters based on the magnetogram-inversion scheme does not use in situ data such as direct counts of electron fluxes or potential drops, but rather relies on indirect observations that measure the magnetic effects caused by electric currents that are carried by charged particles moving under electric fields. Clearly, a realistic model of the height-integrated ionospheric conductivities must be used in the scheme, but no measurements are presently able to sense the instantaneous global distribution of the conductance. The recent availability of radar and satellite data of some ionospheric parameters at certain points and areas, however, tends to provide us with a unique opportunity to make an optimized estimate of the electrodynamic features. It is extremely useful to improve the numerical schemes in such a way that simultaneous, more direct measurements of electric fields, conductivities, and field-aligned currents can be incorporated into the magnetogram-inversion algorithms.

The algorithm has now been improved in several ways: first, the component of the auroral enhancement of the ionospheric conductance has been calculated on the basis of global images of auroral emissions observed from 20000-km altitude with the Dynamics Explorer (DE) 1 satellite (Frank et al. 1982, 1986), instead of employing the statistical models that were used extensively in earlier studies (Kamide et al. 1986). Although the absolute value of the estimated conductivity needs further cross-calibrations with calculations relating to the excita-

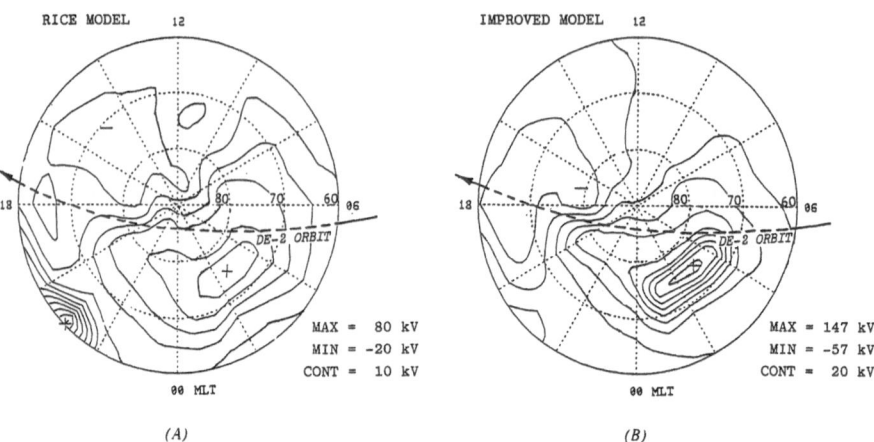

Fig. 4.15. a Comparison of the electric potentials calculated for two different conductivity models for the maximum epoch of an intense substorm, during which nearly simultaneous DE-1 and DE-2 data were available. The potential at the north pole is normalized to zero. Note that different contour intervals are used for *A* and *B*: *A*, 10 kV; *B*, 20 kV

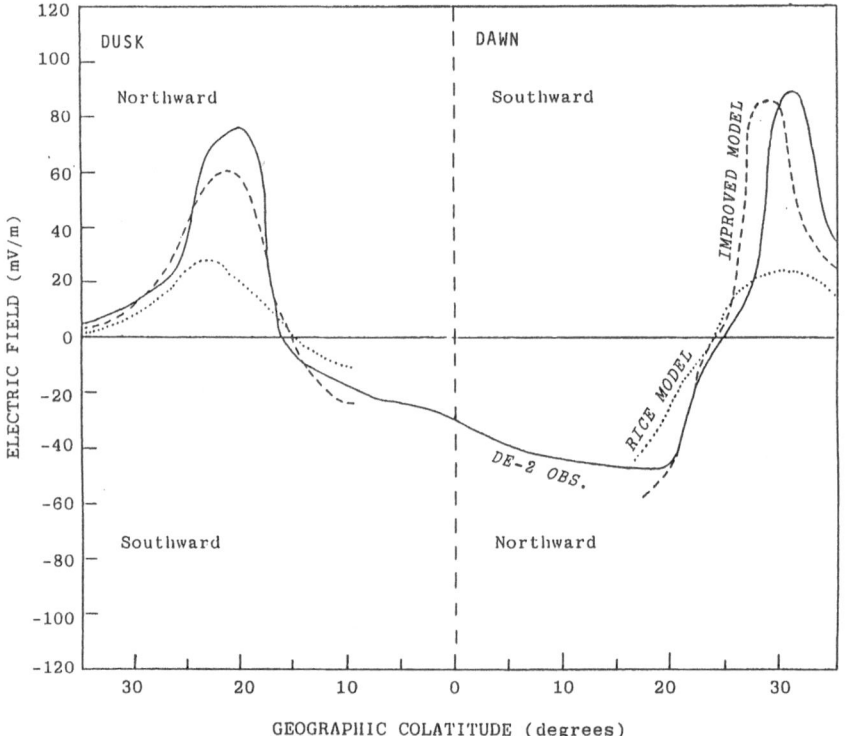

Fig. 4.15. b Observed electric fields approximately in the north-south direction along the DE-2 orbit together with calculated values using two conductivity models. See text for the two models (Kamide et al. 1989)

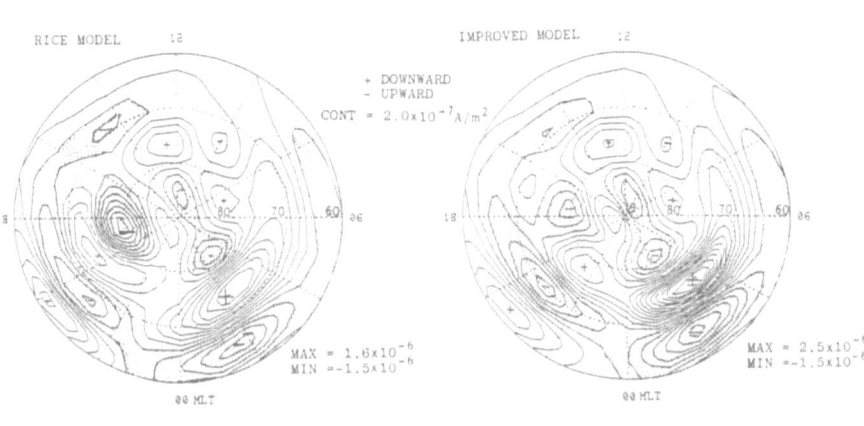

Fig. 4.15. c Comparison of calculated field-aligned current densities for two different conductivity models, (a) the Rice model and (b) the improved model, for the maximum epoch of an intense substorm, during which nearly simultaneous DE-1 and DE-2 data were available. (Kamide et al. 1989)

tion of ionospheric constituents, the region of aurorally enhanced conductivities can be fairly accurately determined from the auroral image data. Further, the use of simultaneous data of ion drifts from the sister satellite, DE 2, at lower altitudes (300 km) makes it possible to check the accuracy of our estimation by comparing the calculated electric fields with ion-drift measurements, although such a comparison can be properly made only along DE 2 orbits (Kamide et al. 1989). The optimum conductivity distribution was then chosen on an iterative basis, such that the resultant electric fields become consistent with DE 2 ion drifts. For this purpose, the Rice conductivity model was modified. For the maximum epoch of an intense substorm, Fig. 4.15a shows the distribution of the electric potential calculated from ground magnetometer data combined with the optimum conductivity. The potential distribution for the Rice model (Spiro et al. 1982) without modifications is also shown on the left-hand side. The overall potential distributions for the two cases are represented by the familiar two-cell pattern. It is noted that the total potential difference across the polar cap is increased by a factor of two by using the improved conductivity model. While both models are likely to have difficulty in determining the electric potential accurately in the polar cap, where the conductance is very low, the improved model tends to yield less complicated potential patterns.

Figure 4.15b compares the electric field estimated from measured ion drifts along the DE 2 orbit approximately in the north-south direction with the electric field, calculated using the two conductivity models (the Rice model and the improved model). The position of the satellite orbit relative to the overall potential pattern is shown in Fig. 4.15a. It is evident that the locations of the reversals in the sign of the electric field, signalling the polar cap boundary, do not change very much: both cases are in good agreement with satellite observations. However, the use of the statistical Rice model gives a small electric field at auroral latitudes, especially in the dawn sector. On the other hand, the improved conductivity distribution reproduces the magnitude of the DE 2 electric fields (maximum 80–100 mV/m along the orbit) reasonably well at both auroral latitudes and in the polar cap. Note that although the polar cap portion of the calculated electric field is not plotted because of large fluctuations caused by low conductivity values, the averages of the calculated electric field for the improved conductivity agree with the DE 2 values.

Among other numerous ionospheric parameters derived with the magnetogram-inversion scheme, Fig. 4.15c shows the distribution of calculated field-aligned current densities for the two conductivity models. The overall current pattern is dominated by Region 1 and 2 currents, which relate to the auroral electrojets (not shown here). We note that the maximum current is increased by 50–70% when the improved model for conductivity is substituted for the Rice model. It is important to note that despite the significant changes in the adopted conductivity models, there is no major difference between the two diagrams in terms of the global pattern of the field-aligned current. This confirms the earlier finding (Kamide et al. 1981) that the ionospheric and field-aligned current distributions are not very sensitive to the choice of the conductivity model as long as the conductive belt is reasonably well co-located in latitude with enhanced auroral activity. This implies that as long as we deal only with the electric currents, we need not pay too much attention to the determination of an accurate conductivity distribution.

In the second form of improvement, many different types of simultaneous measurements can be used (Richmond and Kamide 1988), such as electric fields and ionospheric conductivities from radar and polar-orbiting satellites, field-aligned currents from satellites and ionospheric currents from radar, as well as ground-based magnetic perturbations (see Richmond and Kamide 1988 for mathematical details of the procedure). This updated technique also makes use of available statistical information about the electrostatic potential in such a way that in regions where any data are lacking, the deduced pattern tends toward the statistical pattern. One of the advantages of the new algorithm is that it can quantify the errors inherent in mapping the electric fields, thus demanding more specific quantitative input. Figure 4.16 outlines the improvement (Richmond and Kamide 1988). The practical purpose of the approach is to determine internally consistent patterns of the electrostatic potential, the ionospheric current, the field-aligned current, and magnetic perturbations at the ground and at satellite heights. In other words, even if point measurements by a satellite require us to modify not only local but also global structures of the ionospheric parameters, we must test the modifications quantitatively in terms of their feasibility. Towards this goal, each of the quantities shown in Fig. 4.16 is expanded in a series of "basis" functions, and a statistical method of optimization is subsequently followed to calculate expected values.

Figure 4.17 shows the results of the application to a case that was previously analyzed by the simple KRM algorithm. This example represents approximately the maximum epoch of the intense substorm with the westward electrojet intruding

IONOSPHERIC ELECTRODYNAMICS MAPPING

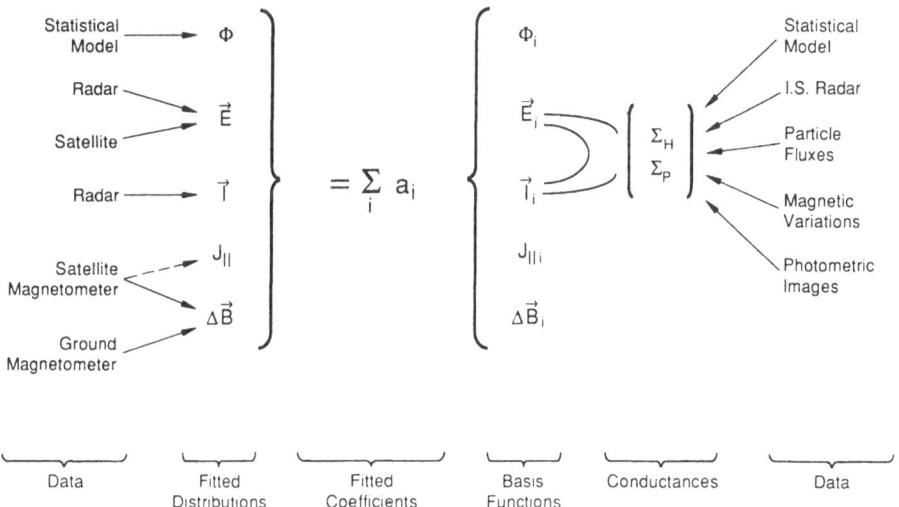

Fig. 4.16. Schematic diagram showing elements of the mapping procedure in the improved magneto-gram-inversion scheme of Richmond and Kamide (1988). There are potentially several types of data about the electric fields, currents, and magnetic variations that can be used to obtain fitted distributions (*left*). Similarly, several types of data may be available to help determine the conductances (*right*). The basis functions for the various electrodynamic parameters are physically interrelated, such that a single set of fitted coefficients is applicable simultaneously to the fitted distributions of all electrodynamic parameters

deeply into evening hours. It is seen that, although the two techniques give generally similar results, there are some noticeable differences, especially for the electric potentials in regions of low conductance. The method after improvement appears to yield less complex electric fields in such regions.

In the third form of improvement, an instantaneous conductivity distribution deduced from bremsstrahlung X-ray image data from the DMSP-F6 spacecraft was combined with the magnetogram-inversion technique (Ahn et al. 1989). The satellite X-ray imagery has a great advantage in that the scanning detector can image a relatively large portion of the auroral belt under both sunlit and non-sunlit conditions. In upgrading the numerical algorithm, the ionospheric con-ductance, calculated from the precipitating electron spectrum on the basis of the X-ray image data following the so-called maximum-entropy method (Gorney et al. 1985), is utilized. Figure 4.18a consists of six polar plots plotted shortly after the maximum phase of a substorm: two of them represent input to the magneto-gram-inversion technique and the other four are output. The instantaneous Hall conductance is deduced from the X-ray image data, showing that, besides the overall enhancement along the auroral belt, there is a surge-like enhancement in the midnight sector.

Figure 4.18b shows an example of overlapping plots of the Hall conductance with the electrostatic potential (upper plot) and with ionospheric current vectors.

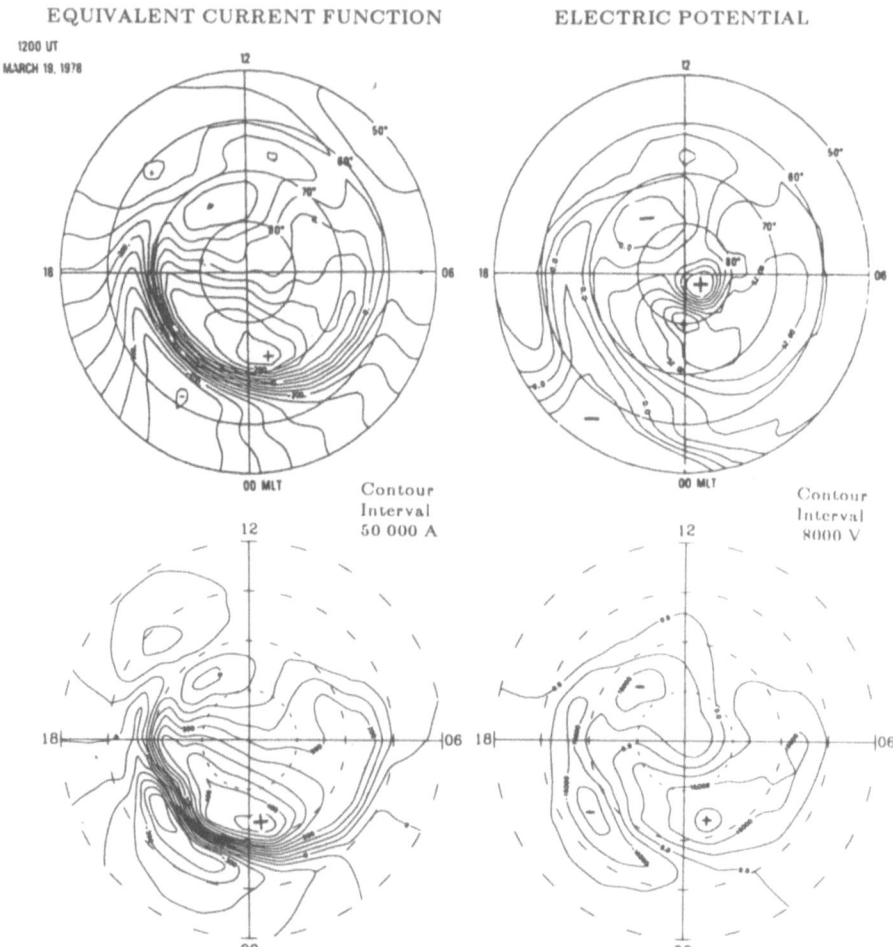

Fig. 4.17. Comparison between the mapping procedure of Kamide et al. (1981) and that of Richmond and Kamide (1988) in terms of the equivalent current function and the electric potential in the ionosphere

It is evident that the equatorward portion of the westward electroject in the morning sector is embedded in an enhanced conductance zone, while its poleward portion seems to be located in a strong electric field region. However, a somewhat different situation is found in the midnight sector, where an enhanced conductance region with no significant accompanying ionospheric current, equatorward of the westward electrojet in the postmidnight quadrant, suggests that an enhancement of conductance even in the nightside auroral latitude does not necessarily accompany an enhanced ionospheric current. On the other hand, except for a patch-like enhancement in the dusk sector, most of the eastward electrojet is located in the region of enhanced conductance of solar UV origin. This indicates that the eastward electrojet seems to be generally dominated by the electric field.

The field-aligned current distribution in Fig. 4.18a also shows well-defined Region 1 and 2 current systems in both the dawn and dusk hemisphere, except that the downward Region 2 current is confined within the postnoon quadrant. Only the upward Region 2 current in the morning sector is reasonably well matched with the enhanced conductance zone, indicating that the electron spectrum is hard enough to produce a significant conductance enhancement in the region.

The isocontours of the Joule heating rate in Fig. 4.18a show that the major heat production regions roughly delineate the auroral electrojets. However, a closer look reveals that the major heating region in the morning sector is located along the poleward portion of the westward electrojet, thus indicating that the electric field is more important than the conductance in the poleward portion of the westward electrojet and the opposite trend prevails in its equatorward portion. On the other hand, although the evidence is less certain, the tendency seems to be reversed in the eastward electrojet region. As expected, the local midnight sector, where the conductance enhancement seems to be more important than the electric field, does not show any significant Joule dissipation.

4.3 Formation of Auroral Arcs

Auroras are upper atmospheric emissions of visible light, resulting from precipitation of primary 1–10 keV electrons. There are basically two types of auroras, diffuse and discrete, into which auroras are morphologically classified (Akasofu 1976). The discrete auroras, including bright, active auroral forms such as auroral arcs, are associated with electrons distributed over a relatively narrow energy range near several keV (e.g. Evans 1968). The angular distributions of these electrons are found to peak in the downward direction along magnetic field lines (e.g. Arnoldy et al. 1974; Mizera et al. 1976), indicating the importance of acceleration by electric fields aligned parallel to the field lines in auroral arcs. The same fields can accelerate ionospheric ions upward (e.g. Shelley et al. 1976; Mizera and Fennell 1977; Chiu and Schulz 1978). On the other hand, the electron precipitation over the diffuse aurora is distributed over a wide range of energies, without being accelerated by field-aligned electric fields.

The interaction of the magnetosphere and the high-latitude ionosphere is a time-dependent, highly non-linear process. Particle precipitation, which carries field-aligned currents, can change the ionospheric conductivity and thus the ionospheric current distribution. Since the ionospheric current must close via the field-aligned currents, the increase or decrease in the auroral current system leads to a feedback "closed" process. In other words, a self-consistent, time-dependent treatment among the auroral electrojets, the field-aligned currents, conductivity enhancements, and the associated electric field must be followed to understand how auroral arcs are formed in the framework of magnetosphere–ionosphere coupling. The interaction (or feedback) between field-aligned currents carried by hydromagnetic waves and increased ionization caused by these currents generate an unstable wave in the polar ionosphere.

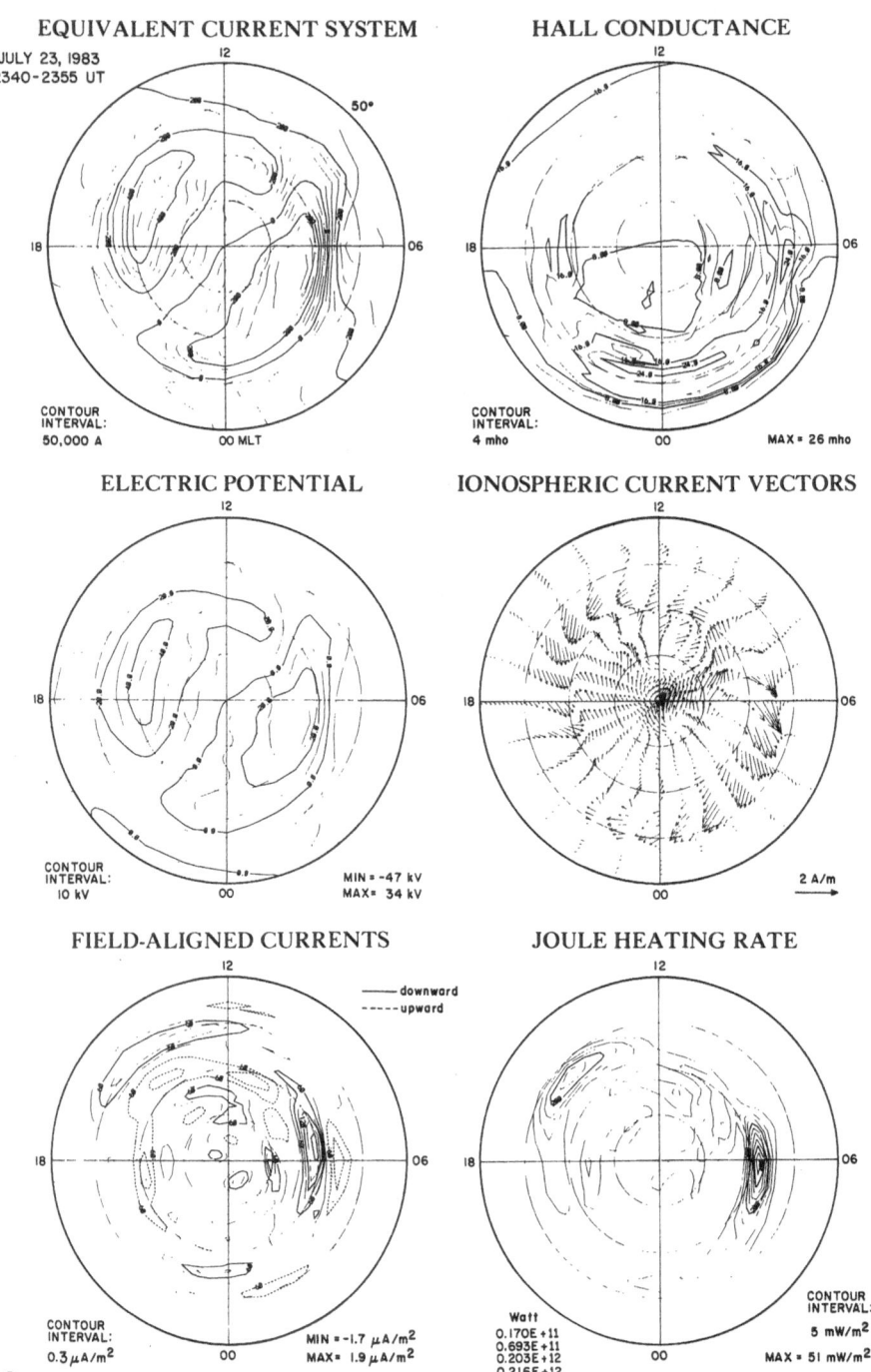

a

Fig. 4.18a

ELECTRIC POTENTIAL AND HALL CONDUCTANCE

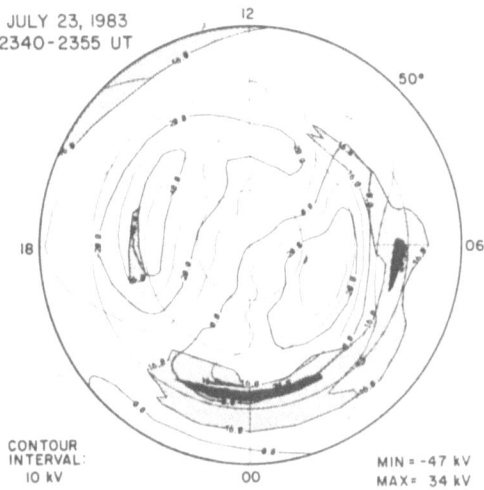

IONOSPHERIC CURRENT VECTORS AND HALL CONDUCTANCE

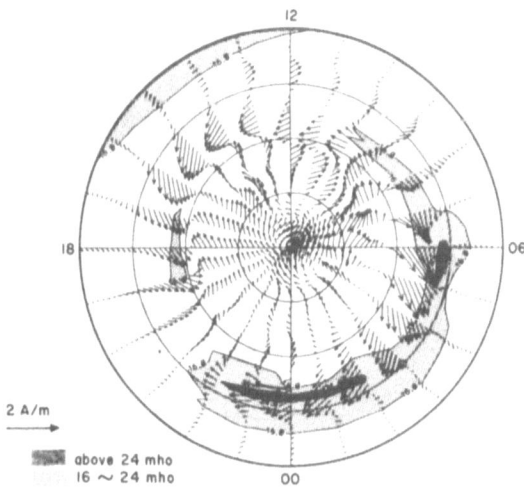

Fig. 4.18. a The equivalent current, Hall conductance, electric potential, ionospheric current vectors, field-aligned currents and Joule heating rate distributions on July 23, 1983. The min and max values listed in the *lower right corner* of the field-aligned current distribution stand for the maximum upward and downward current densities, respectively. The four numerical values in the *lower left corner* of the Joule heating rate distribution denote the globally integrated heating rate from the pole to 80°, 70°, 60°, and 50° in latitude. The maximum Joule heating rate is also shown in the *bottom right corner* (after Ahn et al. 1989). **b** The spatial relationships between the electric potential distributions and enhanced Hall conductance regions (*above*) and the relationship between the ionospheric current vectors and enhanced Hall conductance regions (*below*). (After Ahn et al. 1989)

This coupling was recognized by Sato (1978), who used a lumped circuit in which the magnetospheric response to the ionospheric conductivity was governed by an assumed characteristic impedance in the magnetosphere, and by Rothwell et al. (1964), who used an assumed parameter expressing the efficiency of discharge of excessive charges (or the ionospheric conductivity) as field-aligned currents. In these attempts, the magnetospheric response was assumed to be given by a constant impedance, ignoring the inhomogeneity of the magnetosphere. Lysak (1986) investigated the feedback between the magnetosphere and the ionosphere by means of a model of an Alfvén wave described by Lysak and Dum (1983). This model assumes that an Alfvén wave set up in the ionosphere to carry field-aligned currents will undergo partial reflections as it propagates toward higher altitudes in the magnetosphere, because the Alfvén speed varies along the field line. In any event, the so-called feedback instability (Kindel and Kennell 1971) can take place in the magnetosphere–ionosphere system.

Miura and Sato (1980, 1981) performed detailed simulation studies in which non-linear, coupled equations specifying enhancements in the ionospheric conductivity, field-aligned currents, and the characteristic impedance in the

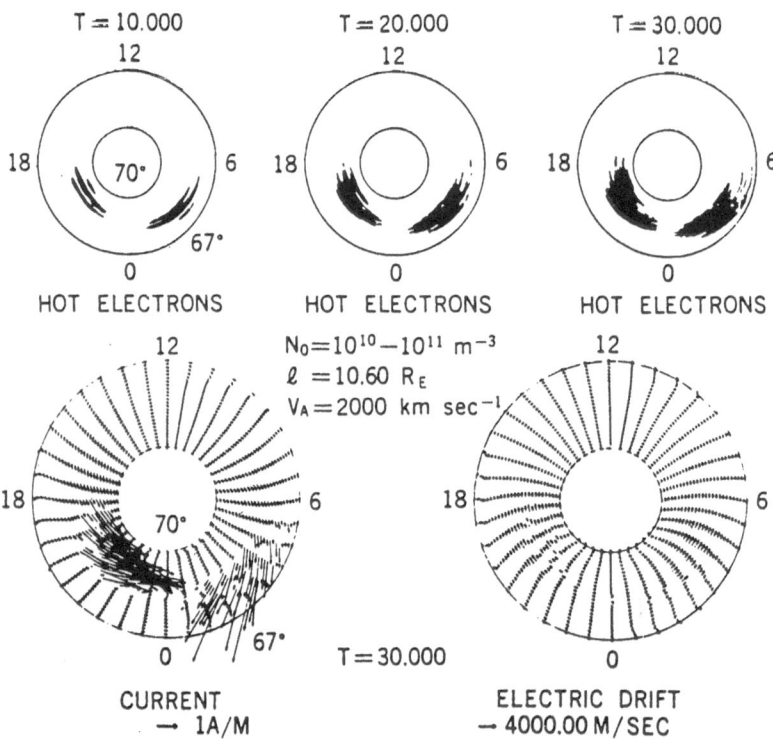

Fig. 4.19. Development of auroral arcs in terms of the auroral electron fluxes in the growing phase for $T = 10$, 20 and 30 (*top row*), ionospheric current vectors for $T = 30$ (*bottom left*), and electric drifts for $T = 30$ (*bottom right*). (After Miura and Sato 1980)

magnetosphere are solved. As time passes in the simulation scheme, the iono-sphere–magnetosphere feedback instability increases, resulting in the precipita-tion of accelerated electrons coincident with upward, field-aligned currents. In the simulation, the initial "background" configuration has been specified to re-present the empirical distribution of the large-scale, field-aligned current system and the potential drop across the polar cap. Figure 4.19 shows an example of their numerical results. The upper three polar plots show temporal developments of multiple auroral arcs which were generated on the poleward side of the premid-night auroral oval and on the equatorward side of the postmidnight oval. The two diagrams in the lower panel show the distributions of the ionospheric current vectors and the electric drift for T = 30. Although such a small-scale structure has been difficult to observe (for examples of fine-scale fields and currents associated with the small-scale auroral arcs, see Sugiura et al. 1982 and Weimer et al. 1985), the feedback instability processes appear to be able to produce such a structure.

The motion of auroral arcs can also be addressed in computer simulation studies in terms of the ionospheric feedback processes. Lysak (1986) showed that a poleward motion of auroral arcs can be reproduced, corresponding to the poleward expansion during the expansion phase of auroral substorms. In parti-cular, although Sato (1978) and Miura and Sato (1980) applied the feedback model to quiet auroral arcs, Lysak (1986) demonstrated that the dynamic motion of auroras, reaching approximately 10 km/s, can also be accounted for by the feedback model.

5 Current Issues of Magnetosphere–Ionosphere Coupling

Our ultimate goal is to understand the entire chain of the magnetosphere–ionosphere coupling self-consistently, as well as to understand each link individually. Our movement toward this goal may be slow, because observations, which are bounded by a number of limitations, such as instruments, orbits, data sampling, etc. give us only a glance of the whole picture, and also because theory and modeling studies are based on many simplifying assumptions that must be tested against the observations. This chapter attempts to highlight recent major progress made in studies of magnetosphere–ionosphere coupling using a wide variety of techniques. Emphasis is placed on pointing out the key issues in the area that need substantial innovation in the near future. This includes the importance of time-dependent treatment of the entire coupling system and the problem of differentiating "disturbed" electrodynamic processes during substorm expansion and those associated with enhanced convection.

5.1 The Westward Traveling Surge

In recent years, significant progress has been made in the study of the three-dimensional current system associated with the westward traveling surge (WTS; see Sect. 2.2.2). When discussing the WTS, it is important to realize that the traveling surge is more than just the surge-like structure traveling westward during substorms and it is not merely one of the many different auroral forms. In fact, the WTS is the most important region in understanding substorm dynamics, because it is the region where the auroral breakup takes place when a magnetospheric substorm starts, and also where the aurora is brightest and the field-aligned current is most intense. The WTS has also been identified as the source of Pi2 pulsations (Pashin et al. 1982; Samson and Rostoker 1983; Lester et al. 1984). The WTS is, in simplest terms, the western end of the westward electrojet as it expands progressively westward starting near local midnight at the onset of substorm expansion. Thus, the Harang discontinuity and the so-called substorm current wedge form in this region. The shape of the WTS is similar to that of giant ocean waves. A typical surge speed is 1–2 km/s. The energy of precipitating electrons associated with the WTS is in the keV range, with the energy depending on the location relative to the surge's head. Note that recent observations of individual "hot spots" in satellite auroral images indicate that the formation and the motion of a traveling surge are more complex than previously thought (Rostoker et al. 1987b). Understanding this region and its relation to the onset of substorm expansion is an outstanding research problem (Kan et al. 1988; McPherron 1991).

5.1.1 Dynamics

Efforts have been made recently to develop a dynamic model of the WTS. Rothwell et al. (1984) proposed a unified model for propagation of the WTS that can explain the diversity in the observed WTS characteristics. They started with the model of the Cowling mechanism as shown in Fig. 2.22 (Inhester et al. 1981; Baumjohann, 1983). The essence of this dynamic model is that precipitating electrons at the conductivity gradient modify that gradient, causing the region of the gradient to propagate as a wave front.

The model starts with a slab in which the westward electric field E (in Fig. 2.22) drives a northward Hall current ($J_H = \Sigma_H E_0$). However, the net ionospheric current may be unequal to J_H, depending on the presence of polarization charges at the northern slab boundary. Since these polarization charges produce a southward electric field and thus a southward Pedersen current, the net current reaching the boundary can be written as

$$J_n = \Sigma_H E_0 - \Sigma_P E_P$$
$$= \alpha J_H,$$

where α is a measure of the polarization efficiency. Here, α is a coupling parameter, expressing also the degree to which the net ionospheric current is closed into the magnetosphere via field-aligned currents. That is, the $\alpha = 1$ situation implies the full connection, and the $\alpha = 0$ value denotes no closure. In the model, α can become > 1, in which case negative polarization charges are accumulated at the northern boundary by excess field-aligned currents, i.e. overclosure.

The time rate of change in the Hall conductivity $\Sigma_H (\approx eN/B)$ is given by

$$\frac{\partial \Sigma_H}{\partial t} = \frac{eh}{B} \frac{\partial N}{\partial t}, \tag{5.1}$$

where N is the ionization density that is related to the precipitating electrons through the ionization efficiency Q (see Rees 1963), h is the appropriate height interval for Q, and B is the magnetic field. The continuity equation for the ion density can be written as

$$\frac{\partial N}{\partial t} = (\text{production rate}) - (\text{recombination rate}) = Q j_{\parallel}/e - \sigma_r N^2, \tag{5.2}$$

where $j_{\parallel}$ is the field-aligned current density and σ_r is the electron-ion recombination coefficient in the ionosphere. Thus, the divergence of the Hall current on the poleward boundary relates to the field-aligned current as

$$j_{\parallel} = -\alpha \frac{\partial J_n}{\partial x} = \frac{\alpha \partial(\Sigma_H E)}{\partial x} = -\alpha E \frac{\partial \Sigma_H}{\partial x}, \tag{5.3}$$

where x denotes the northward direction. If one can momentarily neglect the recombination rate term, Eq. (5.1) becomes

$$\frac{\partial \Sigma_H}{\partial t} = -\frac{QhE\alpha}{B} \frac{\partial \Sigma_H}{\partial x} \quad (\text{for } j_{\parallel} Q/e \gg \sigma_r N^2), \tag{5.4}$$

which is a wave equation with a phase velocity $V_n = Qh\alpha E/B$ in the northward direction. This velocity can also be written as

$$V_n = QhV_d\alpha,$$

where h is the ionospheric height over which Q is significant and the drift velocity V_d is

$$V_d = E_0/B \sim 0.25 \text{ km/s}.$$

After repeating a similar but slightly different treatment for the westward direction, the resulting phase velocity in the northwestward direction can be obtained by combining these two, i.e. $V^2 = V_n^2 + V_w^2$. For the westward direction, a smaller α implies a larger E_P which drives a westward Hall current that adds to the original Pedersen current driven by E_0, as

$$J_w = \Sigma_P E_0 + \Sigma_H E_P, \quad \text{where } J_P = \Sigma_P E_P = (1-\alpha)J_H = (1-\alpha)\Sigma_H E_0.$$

The resulting phase velocity in the westward direction is $V_w = QhV_d[1 + R^2(-\alpha)]/R$.

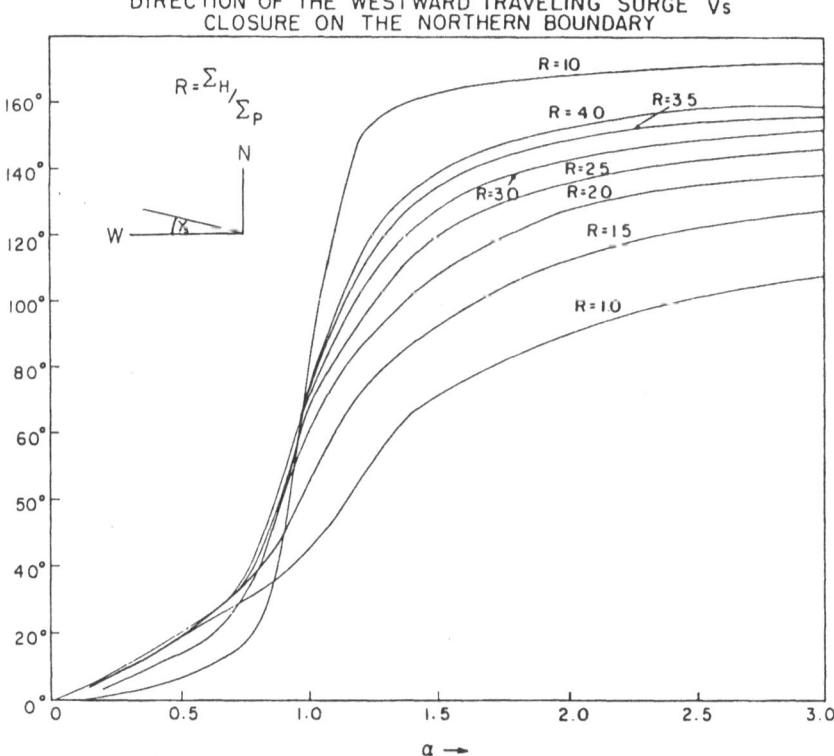

Fig. 5.1. The direction of the surge motion as a function of the closure parameter α in a WTS model. The sensitively of the surge direction to α is dependent on the ratio of the Hall to Pedersen conductivities at the western surge boundary, which is denoted by R. Note that $\gamma_s \sim 45°$ corresponds to $\alpha \sim 0.8$–0.9. (Rothwell et al. 1984)

The total surge speed is known to be much larger than the simple $\mathbf{E} \times \mathbf{B}$ speed and is in quite a different direction.

Figure 5.1 shows the results of the numerical calculation in terms of the direction of the surge motion. The direction γ_s is defined as

$$\gamma_s = \tan^{-1} V_n/V_w$$

It is seen that, although the surge direction can range from due west ($\gamma_s = 0$) to northeast ($\gamma_s > 90°$), the most realistic direction, i.e. northwestward, corresponds to $\alpha < 1$, in particular, $\gamma_s = 45°$ corresponds to $\alpha \sim 0.8 - 0.9$, depending on the ratio between the Hall and Pedersen conductivities.

This model, put forward by Rothwell et al. (1984), predicts that the motion of the WTS is controlled by at least three factors: (1) the energy and intensity of precipitating electrons, (2) the Hall to Pedersen conductivity ratio, and (3) the degree of ionospheric current closure into the magnetosphere. For example, the model suggests that when the incident electron energy changes from 1 to 10 keV, the surge speed should increase from 2 to 34 km/s. It is, however, important to point out that this dynamic model of the WTS includes the dynamics only of the ionosphere. A future task is to properly treat the magnetospheric dynamics as well. It is also important to realize that the ionospheric conductivity does not, in general, remain stationary in the WTS moving frame (see Rothwell et al. 1986, 1988 for the effects of a non-stationary conductivity profile).

5.1.2 Distortion of Convection Pattern

Kan et al. (1984) proposed a mechanism for the distortion of the large-scale convection pattern associated with the westward traveling surge. The basic idea of the proposed mechanism is the partial blockage of field-aligned currents associated with the divergence of the Hall current, resulting in the generation of the polarization electric field. This idea is a generalization of the well-known Cowling conductivity concept for the auroral electrojet (e.g. Boström 1974) introduced for the WTS by Baumjohann and co-workers (see Sect. 2.2.2).

Kan and Kamide (1985) modeled the WTS electrodynamics in the framework of the global convection pattern. In earlier models of the kind (Kan et al. 1984),

▶

Fig. 5.2. a (*A*) Equipotential contours of the input potential Φ_0 in the model of Kan and Kamide (1985). The total potential difference across the polar cap is 100 kV, which is given by the difference between the maximum and minimum potential indicated in the *lower lefthand corner*. The potential difference between neighbouring contours is 5 kV. (*B*) Contours of the Hall conductivity with 2-S difference per contour. The maximum conductivity of 15 S occurs at midnight between 60° and 70° latitudes. **b** (*A*) Equipotential contours (at 5-kV intervals) of the total potential distorted by the blockage process including the conductivity enhancement effect. The polar cap potential difference is 67.27 kV. (*B*) Ionospheric current vector produced by the potential in (*A*) and the conductivity in **c**. The scale of the current is shown in the *left-hand corner* by the vector (0.5 A/m). **c** (*A*) Contours (at 4-S intervals) of the Hall conductivity enhanced by the upward field-aligned current. The maximum conductivity occurs at 1800 MLT. (*B*) Contours of constant field-aligned current density at $0.1\,\mu\text{A/m}^2$ intervals. Upward current is denoted by the *dashed curves*, whereas downward current is shown by the *solid curves*

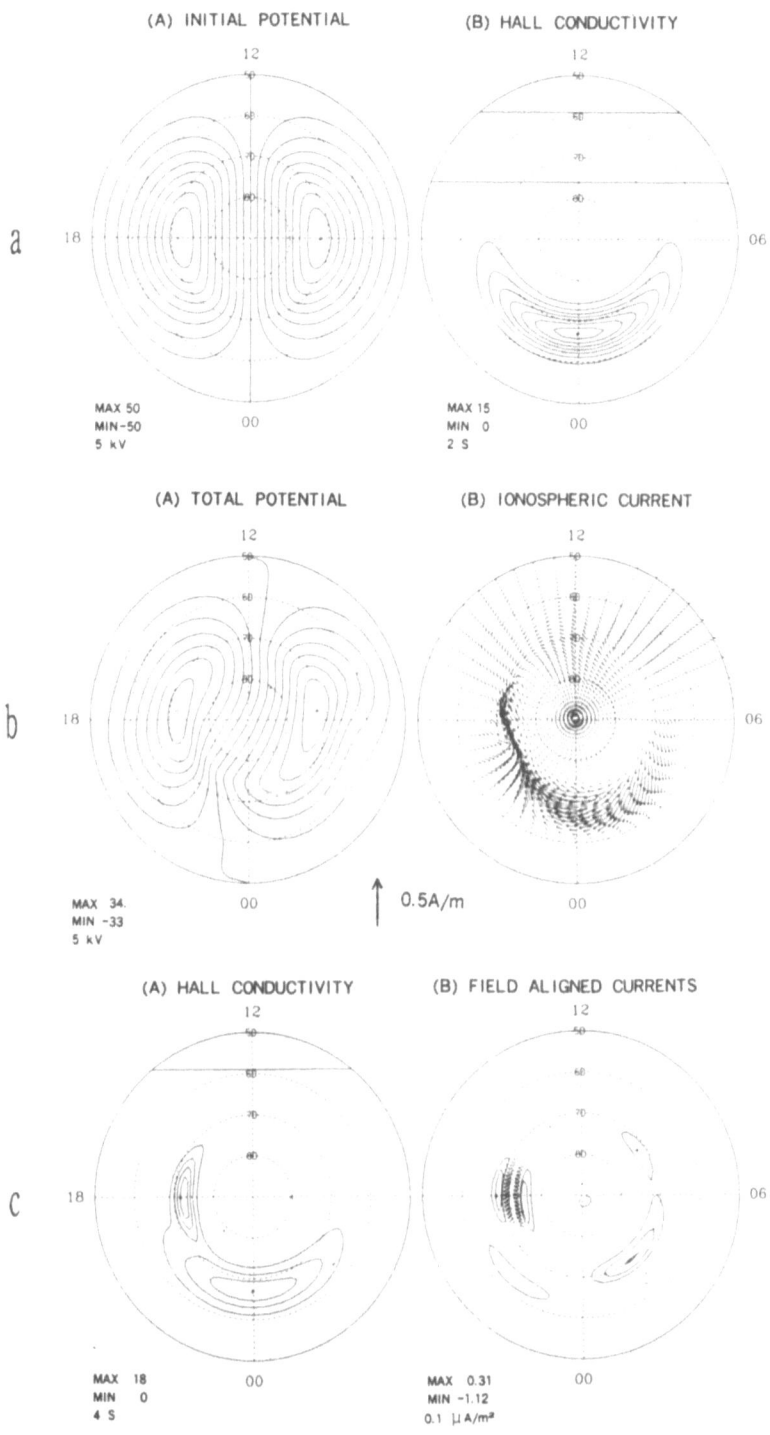

several important parameters in the ionosphere were given independently and fixed throughout the calculation. In the Kan and Kamide model, however, the ionospheric conductivity is allowed to increase self-consistently with increasing upward field-aligned current. Figure 5.2a shows the input potential and conductivity distribution. After repeating computer calculations until convergence is reached, Fig. 5.2b,c is obtained. In the process of the calculation (enhanced conductivity leads to upward field-aligned current, which in turn leads to a new conductivity structure), the WTS region moves westward. Note, however, that this is not a time-dependent model, but that the inclusion of the mutually coupled "conductivity and field-aligned current" is intended to generate a localized, intense westward electrojet on the poleward side of the Harang discontinuity, moving westward with the progress of iteration, not with the progress of actual time. It is also found that the westward electrojet rotates counterclockwise around the leading edge of the surge, consistent with previous observations.

Blockage of the divergence of the Hall current can be written as

$$\nabla \cdot [\Sigma_p \mathbf{E}_p + \Sigma_H \hat{\mathbf{B}}_0 \times (\mathbf{E}_0 + \mathbf{E}_p)] = 0, \tag{5.5}$$

where $\mathbf{E}_0$ is the electric field externally impressed on the ionosphere, and $\mathbf{E}_p$ is the polarization electric field produced by the blockage. This equation states that the divergence of the Hall current of the total electric field $\mathbf{E}(= \mathbf{E}_0 + \mathbf{E}_p)$ is counter-balanced by the divergence of the Pedersen current of $\mathbf{E}_p$.

In addition to blocking the divergence of the Hall current, it is conceivable that the divergence of the Pedersen current can also be blocked. If the Pedersen current is blocked, then the electric field parallel to $\mathbf{E}_0$ will be reduced in magnitude. In the model, Φ_0 is assumed to be a constant-voltage source which automatically excludes the effect of blocking the divergence of the Pedersen current. On the other hand, the polarization field due to the blockage of the Hall current is primarily perpendicular to $\mathbf{E}_0$.

Substituting $\mathbf{E}_p = -\nabla\Phi_p$ and $\mathbf{E}_0 = \nabla\Phi_0$ into Eq. (5.5), one obtains the following second-order equation for the potential Φ_p of the polarization field:

$$\nabla^2\Phi_P + \frac{\nabla\Sigma_P}{\Sigma_P}\cdot\nabla\Phi_P + \frac{\nabla\Sigma_H}{\Sigma_P}\cdot\hat{\mathbf{B}}_0 \times (\nabla\Phi_0 + \nabla\Phi_P) = 0. \tag{5.6}$$

The corresponding convection pattern distorted from the Φ_0 pattern is given by $\Phi(= \Phi_p + \Phi_0) = $ const. The ionospheric current $\mathbf{J}$ and the field-aligned current $j_\parallel$ can then be calculated from

$$\mathbf{J} = \Sigma_P(\mathbf{E}_P + \mathbf{E}_0) + \Sigma_H\hat{\mathbf{B}}_0 \times (\mathbf{E}_P + \mathbf{E}_0) \tag{5.7}$$

and

$$j_\parallel = -\nabla\cdot\mathbf{J}, \tag{5.8}$$

respectively, where $j_\parallel < 0$ (or > 0) for upward (or downward) current flowing away from (or into) the northern hemisphere ionosphere.

To include the self-consistently enhanced conductivity in Eqs. (5.6) and (5.8), Kan and Kamide (1985) related the conductivity to the field-aligned current density, assuming that the field-aligned currents are carried primarily by electrons in the region where $\nabla\cdot\mathbf{E} < 0$ (Lyons 1980). The global patterns of the electric potential shown in Fig. 5.2 are in reasonable agreement with recent observations by means of radar and satellite techniques, in the sense that the deformation of the potential vortex lines occurs in the premidnight sector. The deformation is even more pronounced in the distribution of ionospheric current vectors, because the ionospheric conductivity is strongly intensified in this region. On a global scale, the westward electrojet, starting from the morning sector with an approximately 10° latitudinal width, extends into the premidnight sector where the width decreases drastically and exhibits the observed counterclockwise rotation around the leading edge of the WTS (Inhester et al. 1981; Opgenoorth et al. 1983a; see also Sect. 2.2.2.)

5.1.3 Pulsations

Geomagnetic pulsations and magnetospheric substorms have often been treated as separate topics in the literature. Over the past several years, however, efforts have been made to combine the two phenomena physically, in particular the WTS and pulsation phenomena, which share some common properties, including their spatial localization and propagation velocities near local midnight during the substorm onset (Lester et al. 1984; Gelpi et al. 1987).

As described in Section 2.1.2, the WTS has been identified as the source of Pi2 pulsations (Baumjohann and Glaßmeier 1984; Sun and Kan 1985; Rothwell et al. 1986). In particular, Kan and Sun (1985) treated the WTS and the Pi2 pulsations as a consequence of an enhanced magnetospheric convection in a model of magnetosphere–ionosphere coupling. The essence of the Kan and Sun model is the bunching of Alfvén waves launched by the suddenly enhanced convection in the magnetosphere preceding the onset substorms. The reflection of these Alfvén waves from the ionosphere is analyzed using the height-integrated conductivity which is allowed to be highly non-uniform (see Goertz and Boswell 1979; Lysak and Dum 1983). The reflection of the Alfvén waves from the magnetosphere, or equivalently, the closure of field-aligned currents in the magnetosphere, is treated phenomenologically, depending on whether the field lines are open or closed. On open field lines, the reflection coefficient is assumed to be negative, which leads to an enhancement of field-aligned currents at each reflection and the source region is identifiable as a voltage source. On closed field lines, the reflection coefficient is positive, which leads to a blockage of additional field-aligned currents and the source region is identifiable as a current source. Their simulation results, including the convection pattern, the field-aligned current, the conductivity enhancement, the oscillation of the westward electrojet and the average speed of the WTS, are all in reasonable agreement with the features of the WTS and the Pi2 pulsations during the substorms, although the reflection of Alfvén waves from the magnetosphere needs to be studied microscopically from the wave standpoint, in relation to the generation mechanism of field-aligned currents in the magnetosphere (see Cao and Kan 1990).

In the Rothwell et al. (1986, 1988) model, perturbations in the ionospheric current in the WTS region are shown to produce standing waves due to the reflection from conductivity gradients along the surge boundary. Their approach is to model the ionospheric response to precipitating electrons, rather than first determining how the initial Alfvén waves transmit the information on the tail current disruption at the creation of the substorm current wedge. This approach has the advantage of dealing with a tractable portion of the non-linear magnetosphere–ionosphere coupling problem whose plausible solution imposes conditions that the far less tractable magnetospheric source must satisfy.

5.2 Auroral Particle Acceleration and Parallel Electric Fields

A great deal of theoretical work has yielded remarkable success in accounting for observations in the magnetosphere and the magnetosphere under steady-state conditions, by assuming that electric fields $\mathbf{E}$ parallel to a magnetic field $\mathbf{B}$ are zero or very small, i.e. $\mathbf{E} \cdot \mathbf{B}/B = E_\parallel \sim 0$. This assumption has been supported by the rationale that the magnetosphere–ionosphere system has sufficient numbers of cold plasma which are able to short out any field component parallel to $\mathbf{B}$. However, the auroral region, in which the $\mathbf{E} \cdot \mathbf{B} = 0$ condition is imposed, is most clearly violated (Schulz 1991).

The existence of parallel electric fields on auroral field lines is essential to the magnetosphere–ionosphere coupling as a whole, as well as to microscopic acceleration processes of auroral particles. The field-aligned currents are associated with the parallel electric fields which accelerate not only auroral electrons into the ionosphere but also ions out of the ionosphere. Arnoldy et al. (1985) reviewed the earlier measurements of field-aligned electrons and reassessed them in the light of more recent and comprehensive data available from both rocket and satellite observations. One of the main points is that field-aligned electrons are a common feature in evening, midnight and cusp auroras.

5.2.1 Observations

Hallinan et al. (1985) reviewed 13 years of their video-recording studies and reached the conclusion that the spectrum of "enhanced" auroras is grossly similar to that of "background" auroras, implying that auroral enhancement is a result of a localized plasma instability that energizes ambient electrons at the expense of the kinetic energy of the precipitating electrons. There is still the possibility that field-aligned currents are carried directly by "hot" electrons that are accelerated by the reconnection electric field in the tail (Sonnerup 1971; Sato et al. 1982), but it is likely that "colder" electrons are responsible for carrying the field-aligned currents. Moreover, auroral electrons are accelerated through the so-called parallel electric field (see, for example, reviews by Block 1972, 1975, 1978; Evans 1975; Block and Fälthammar 1976; Shawhan et al. 1978; Goertz 1979; Mozer et al. 1980; Fennell 1984; Kan 1984).

Ever since the discovery of the "inverted V" particle spectrum pattern (Frank and Ackerson 1971), parallel electric fields above the auroral altitude have been invoked. In fact, the existence of a kV potential drop along magnetic field lines at auroral latitudes is strongly supported by many different types of observations.

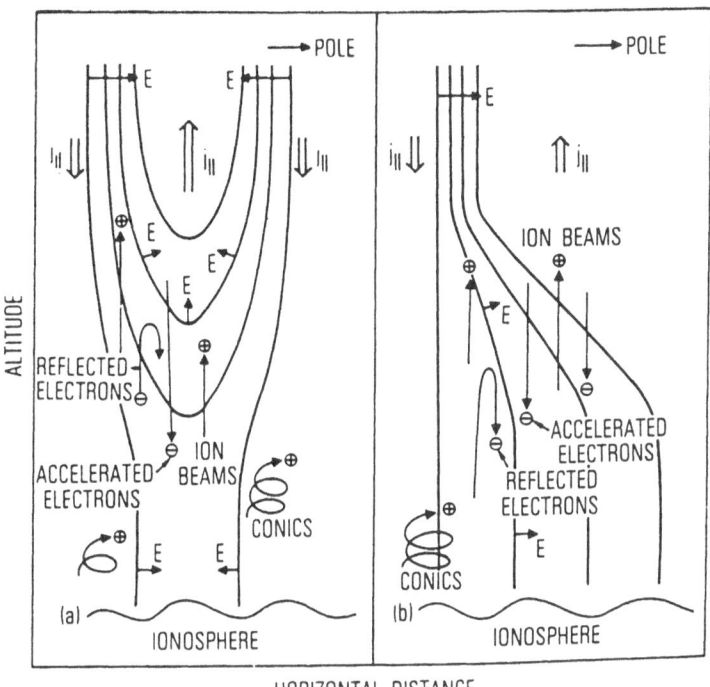

HORIZONTAL DISTANCE

Fig. 5.3. Illustration of V-shaped (*left*) and S-shaped (*right*) electric potential structures. The kinds of particle distributions which are expected to be seen near and in the structures are also shown. (Fennell 1984)

For example, the electron pitch-angle distribution over discrete auroras (Lui and Anger 1973; Akasofu 1974) shows evidence that the angular distribution of these auroral particles is at times peaked along magnetic field lines, indicating that these electrons have been accelerated through a parallel electric field (e.g. Hoffman and Evans 1968; Paschmann et al. 1972; Lundin 1976; Chiu et al. 1983). Further electric field data from polar-orbiting satellites show the signature of electrostatic shocks or double layers confined to altitudes of less than a few R_E (e.g. Mozer et al. 1977; Temerin et al. 1981; Burke et al. 1984). Sometimes these shock structures show paired, oppositely directed electric fields called V-shaped shocks, and at other times they consist of electric fields of one sign, called S-shaped shocks (Mozer 1981; Mizera et al. 1982). Figure 5.3 illustrates the structure of electric equipotentials for these two types along with the types of expected particle distributions (Fennell 1984). The shocks typically span a latitudinal width of 0.01° to 0.1°.

There has recently been an ever-growing body of observational evidence that auroral particles are accelerated at relatively low altitudes ($\sim 1R_E$) along magnetic field lines (e.g. Ghielmetti et al. 1978). Kaufmann and Kintner (1984) used S3–3 measurements of upward moving ion beams as well as precipitating electrons, and found that ions carry most of the parallel momentum flux while electrons carry most of the energy flux within the ion beam regions, implying that these energetic electrons carry field-aligned currents all the way from the ionosphere to the distant tail. Figure 5.4 shows an example of such S3–3 measurements, where it is noted that the three periods marked by heavy dots were the best examples of steady ion beams that they could find. Currents deduced from magnetometer and energetic electron data agree, at least qualitatively, whenever the magnetometer records a significant upward current. Both detectors find that there is a net upward current within ion beams and that the observed ion beams carry only a small fraction of this net current. Data from electrostatic analyzers on board the S3–3 show that ion conics can be trapped at low altitudes by

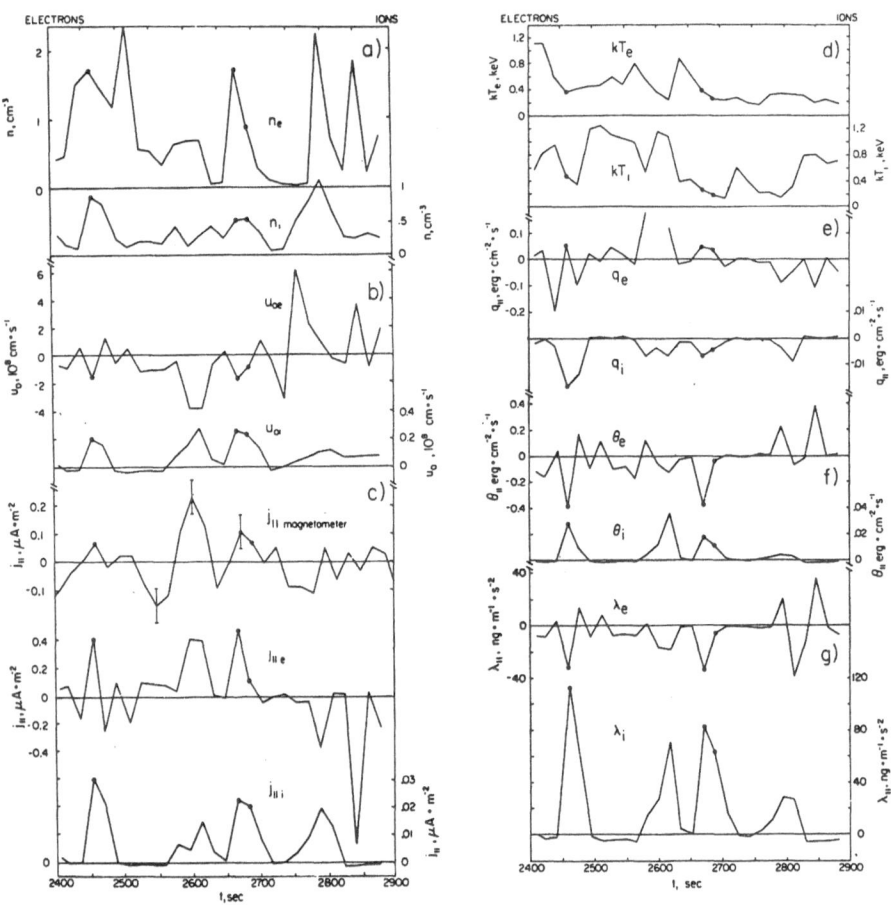

Fig. 5.4. Fluid parameters calculated from ion and electron data from the S3-3 satellite. Each point is based on data from the one full spin period (18 s). Parameters plotted are number density, drift velocity, parallel electric current, temperature, parallel heat flux, parallel energy flux, and parallel momentum flux. The three points marked with *heavy dots* in each panel indicate spin periods during which ion beams were seen. Positive quantities indicate upgoing fluxes or currents. (Kaufmann and Kintner 1984)

downward electric fields (Gorney et al. 1985). It is interesting to note that DE 1 plasma instrument data show electron conic signatures at high latitudes (Menietti and Burch 1985).

In the late 1970s cold ions of ionospheric origin accelerated upward were discovered (see Gorney 1987; Horwitz 1987 for review). Klumpar et al. (1984), Moore et al. (1985) and others used new instruments for measurements on the DE 1 spacecraft. In particular, Klumpar et al. showed evidence of a two-stage acceleration of ionospheric ions in that ions can be accelerated by a mechanism acting primarily transverse to the magnetic field at an altitude near 18 000 km and additionally accelerated through a parallel potential in the core of the inverted V. The ionospheric source of the large abundances of O^+ ions has been actively examined (e.g., Lockwood 1984; Lockwood et al. 1985; Waite et al. 1985). There have also been continued investigations of the nature and origin of the particle population associated with electron beams injected from rockets into the iono-sphere (Winckler et al. 1984; Arnoldy et al. 1985).

Barium clouds released from a rocket provide a useful means of estimating the electric field structure, not only the perpendicular but also parallel field com-ponents. This was demonstrated by tracking barium clouds that were formed by injecting barium along a magnetic field line (Wescott et al. 1972). Cloud motion of Ba^+ ions, which were being accelerated away from the Earth, was observed to infer a parallel energy gain above auroral forms (Haerendel et al. 1976; Wescott et al. 1976). The gain in parallel velocity of the ions can most readily be accounted for by having parallel electric fields which exist above a certain height.

Barium charges released during a small auroral breakup were used by Stenbaek-Nielsen et al. (1984) to study the altitude and the structure of the acceleration region. From a detailed analysis of data, it was concluded that there was no appreciable parallel electric field below 8100 km in this particular experiment and that the barium ions were accelerated upward and dispersed rapidly. The estimated potential was 1 kV distributed over no more than 200 km. It should also be noted that Heelis et al. (1984) showed, using DE 2 measurements, that field-aligned O^+ fluxes of a few $10^8 \, cm^{-2} \, s^{-1}$ are not an uncommon occurrence at most latitudes and local times. This observation confirms the earlier estimate by Frank et al. (1977) and Hultqvist (1983) that an O^+ flux from the ionosphere of $10^8 \, cm^- \, s^{-1}$ is required to account for the magnetospheric population. More-over, Heelis et al. (1984) showed that upward O^+ fluxes in excess of $10^{10} \, cm^2 \, s^{-1}$ can sometimes be observed at 1000-km altitude, carrying very intense $(100 \, \mu A/m^2)$ upward field-aligned currents. Finally, we note that Lin et al. (1984) observed, on the basis of DE-1 measurements, that counterstreaming electrons are generally detected at energies below a few hundred eV in association with energetic (> 1 keV) precipitating electron fluxes (see Wagner et al. 1985 for results of com-puter simulations of the production of counterstreaming electrons associated with parallel field). Yau et al. (1985) also examined energetic auroral ion flow at DE 1 altitude and concluded that energetic (0.01–17 keV/e), upward flowing ions are possibly an adequate source of O^+ in the plasma sheet.

The auroral current–voltage relationship between field-aligned currents ($j_\parallel$) and the parallel potential drop ($V_\parallel$) (e.g. Knight 1973; Chiu and Cornwall 1980; Fridman and Lemaire 1980; Lyons 1980; Yeh and Hill 1981) can be tested by using simultaneous observations of electric fields from a pair of spacecraft (the DE-1/2 pair) along auroral field lines. Weimer et al. (1985, 1987) showed that, within certain constraints, the data support a linear relationship between $j_\parallel$ and $V_\parallel$ (kinetic Ohm's law), and characteristic scale sizes of auroras play a role in determining the field-aligned conductance. The form in individual cases (e.g. Burch et al. 1990; Lu et al. 1991; Marshall et al. 1991), the spatial distribution of the parallel conductance, the ratio of $j_\parallel/V_\parallel$, is a crucial parameter in local magnetosphere–ionosphere coupling, particularly in the acceleration regions.

It is also very important to point out that there are several observational characteristics of the aurora which deserve special attention in auroral theories but for which attempts at formulation have not been made (Haerendel 1983). They are, for example (1) multiplicity of auroral arcs and (2) large electric field, transverse, near the arc, not in the center of the arc. Implicit in most of the auroral arc theories is the existence of intense upward field-aligned currents within the latitudinal span of auroral arcs. The most intense field-aligned currents are, however, often observed at the edges of the acceleration regions.

5.2.2 Theories and Computer Simulations

Several mechanisms have been proposed that could lead to the generation and the sustentation of quasi-stationary, non-zero parallel electric fields. Some models seek plasma turbulence to decrease the parallel conductivity or to impede higher current levels (Kindel and Kennel 1971). Wave–particle interaction results in a non-vanishing, effective parallel conductivity. Hultqvist (1971) suggested that the thermoelectric effect in regions of contact between hot magnetospheric plasma and cold ionospheric plasma could produce an appreciable field-aligned potential drop. As a result of the flux imbalance, net negative charges tend to build up at the top side of the ionosphere. The magnetic mirror force on the current carrying electrons of upward field-aligned current can prevent those electrons from reaching the ionosphere, generating potential drops above the auroral altitude (Knight 1973; Lemaire and Scherer 1974; Lennartsson 1976).

Since it is unlikely that the pitch-angle diffusion rates for magnetospheric ions and electrons are the same, space charge separations tend to develop in the vicinity of the mirror points. To maintain the plasma in a state of quasi-neutrality, an electric field with a significant component parallel to the magnetic field is required (Alfvén and Fälthammar 1963). Chiu and Schulz (1978) solved numerically for the self-consistent potential distribution by using different plasma species. Lyons (1980, 1981) successfully presented a quantitative explanation for electric potentials along magnetic field lines as well as in the ionosphere by requiring discontinuities with a negative divergence of high-altitude magnetospheric electric field, $\nabla \cdot \mathbf{E} < 0$. There is one class of mechanisms which involves double layers and electrostatic shocks, a discontinuity of the electric field resulting from a tendency of the plasma to maintain the charge neutrality in the regions which adjoin it (Stern 1981). These consist of two oppositely charged layers along the magnetic field.

Table 5.1. Mechanisms for field-aligned potential drops

Processes	$j_\parallel$	Energy source	Remarks
(A) Double Layer	$j_\parallel \neq 0$	Electrical Energy (emf)	Auroral Double Layer process $\simeq$ (A) + (C) + (E) + (B) + (D)
(B) Electrostatics shock	$j_\parallel = 0$	Ion streaming energy (mechanical)	$L_\parallel \gg \lambda_D$ $L_\perp \lesssim \rho_i$ Auroral arcs
(C) Pitch-angle anisotropy	$j_\parallel = 0$	Thermal energy	$L_\perp \gg \rho_i$ Inverted V's
(D) Thermoelectric	$j_\parallel = 0$	Thermal energy	Potential drops along field lines $\sim$ 1 to 10 kV
(E) Anomalous resistivity wave-particle interactions	$j_\parallel \neq 0$	Electron streaming energy	Quasi-static $E_\parallel$
(F) Alfvén waves	$j_\parallel = 0$	Wave energy	Transient or periodic $E_\parallel$

$L_\perp$: perpendicular scale length; $L_\parallel$: parallel scale length; ρ_i: ion gyroradius

There seems to be a convergence of understanding of the processes responsible for the formation of parallel electric fields. The main theoretical issue which must be addressed is the mechanism that sustains potential drops along field lines. Table 5.1 (from Kan 1982, 1984) summarizes different mechanism which can sustain potential drops along field lines self-consistently in a collisionless plasma. The important point to make is that these mechanisms are not mutually exclusive. For example, double layers can accelerate electrons, which can then enhance anomalous resistivity and, although the magnetic mirror force may not be able to support the observed large field strength (> 100 mV/m), it may facilitate the formation of double layers or electrostatic shocks. It seems likely that the double layer is the dominant mechanisms responsible for a relatively stable structure of parallel electric fields, although our understanding is limited regarding the assessment of the relative importance of double layers and shocks. There are also small amplitude double layers moving along magnetic field lines (see Hudson et al. 1983; Bujarbarua and Goswami 1985).

We have witnessed recently important advances in our understanding of three areas concerning auroral particle acceleration/parallel electric fields. The first area is simulation studies, which attempt to determine scale lengths of the V-shaped potential structure or of the double layers. Regarding the perpendicular scale length, several auroral arcs are likely to be packed within an inverted-V structure. The field-aligned scale length of the double-layer potential structure remained an unsolved issue associated with the formation of auroral forms. It seems likely that potential drops distributed along field lines over distances greater than the Debye length, up to $1 R_E$, have been suggested from observations (e.g. Boehm and Mozer 1981), while the existence of localized double layers has also been inferred in plasma observations at auroral latitudes. Borovsky and

Joyce (1983a,b), Wagner and Kan (1985), and Yamamoto and Kan (1985) were concerned with the field-aligned scale length of double layers. They showed, by means of numerical simulations, that the auroral potential structures can be either localized or extended, depending on various conditions, such as back-scattered primary electrons, charge neutrality condition, and field configurations. In particular, Wagner and Kan (1985) demonstrated the importance of the effect of the converging magnetic field on the scale length of the potential structure along auroral field lines. It was shown that a localized double layer tends to evolve into an extended potential structure in such a converging magnetic field configuration. In the presence of backscattered primary electrons and/or mag-netospheric hot ions, Yamamoto and Kan (1985) showed that the potential structures can be either localized or extended, depending on whether or not charge neutrality is maintained along auroral field lines. Hudson et al. (1983) studied solitary waves for auroral double layers, finding the possibility of multiple double layers.

The second area is the consideration of the origin of the auroral acceleration. Tiwari and Rostoker (1984) suggested that substorm-associated intensification of field-aligned current is triggered by the incidence, in the ionosphere, of a large amplitude Alfvén wave, generated in the distant magnetotail, demonstrating that the character of the background plasma and magnetic field is the acceleration region near $1R_E$ that can be ideal for the generation of MHD or electrostatic shocks and that these shocks can lead to the acceleration of ions and electrons. Stasiewicz (1984) gave an account of the origin of the auroral electron spectra in the inverted-V region in terms of a quasi-linear theory of ion acoustic turbulence. Computer simulations based on an anomalous resistivity model have been utilized to understand the collective plasma process (Lin and Rowland 1985). Both the simulations and earlier AE-D observations agree that the accelerated electrons undergo large pitch-angle scattering when the potential drop increases above a certain critical value. These simulation models can be compared with an empirical model of auroral electron precipitation as a function of geomagnetic activity, which has also been constructed by Hardy et al. (1985).

The third kind of studies are complete simulations of ion signatures of auroral acceleration processes. The existence of the upward $E_\parallel$ seems to require a con-comitant beam of upward flowing ions above the acceleration region if there is a significant population of ions in the ionosphere. These ions associated with auroral phenomena have, in fact, now been observed above, say, 3000 km in two forms: (1) field-aligned, upward moving, accelerated ion beams, and (2) trans-versely accelerated ions (ion conics). Theoretical progress and computer simula-tions have provided a number of viable mechanisms to explain the observed features of those ions above discrete auroral forms. For example, Okuda and Ashour-Abdalla (1983) suggested the possible importance of ion cyclotron tur-bulence generated by electron currents of ionospheric origin. Novel mechanisms for the preferential acceleration of O^+ and H^+ have been proposed (e.g. Mitchell and Palmadesso 1983; Horwitz 1984; Temerin and Lysak 1984; Gorney et al. 1985). What auroral electron and ion beams tell us about magnetosphere–ionosphere coupling has been discussed by Kaufman (1984).

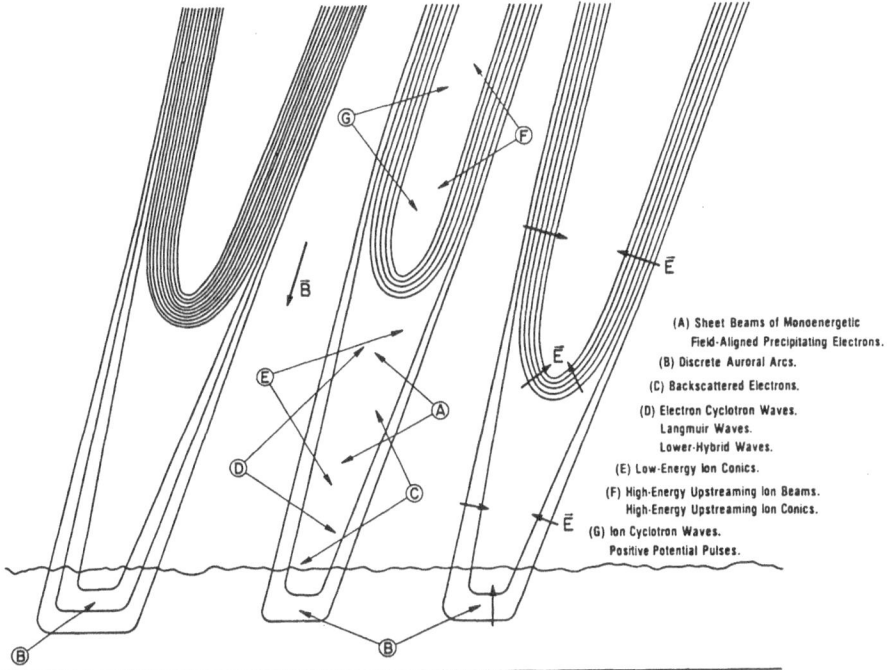

(A) Sheet Beams of Monoenergetic
     Field-Aligned Precipitating Electrons.
(B) Discrete Auroral Arcs.
(C) Backscattered Electrons.
(D) Electron Cyclotron Waves.
     Langmuir Waves.
     Lower-Hybrid Waves.
(E) Low-Energy Ion Conics.
(F) High-Energy Upstreaming Ion Beams.
     High-Energy Upstreaming Ion Conics.
(G) Ion Cyclotron Waves.
     Positive Potential Pulses.

Fig. 5.5. A schematic (not to scale) of the equipotential contours and related phenomena above a multiple auroral arc system as seen in a north-south cut. (Borovsky 1984)

Figure 5.5 is a schematic illustration of two-dimensional potential structure located above auroral arcs in which one can envision electric fields parallel to the terrestrial magnetic field in the central segments while the electric fields become oblique to the magnetic field on either edge. Borovsky and Joyce (1983a,b) and Borovsky (1984) studied the production of ion conics by oblique double layers as well electron beams. For this purpose, they conducted a two-dimensional numerical simulation in which thousands of test ions were inserted in order to determine whether these test ions were subjected to acceleration through an oblique double layer with velocity vectors aligned with or oblique to the ambient terrestrial magnetic field. One of their findings is that accelerated heavy ions will be substantially less field-aligned than accelerated hydrogen ions, a possibility of auroral double layers as a source of ion conics. Greenspan (1984) found that the oblique double layer does increase both the perpendicular and the parallel energy of ions, but the amount of perpendicular energization is greater for O^+ than for H^+, in agreement with observations. A possible mechanism was also reported by Singh and Schunk (1985) to explain the observed streaming of O^+ and H^+ ions at nearly equal speeds in the tail. This mechanism involves the acceleration of the ions through their interaction with electrostatic hydrogen cyclotron waves, on auroral field lines.

5.3 Penetration of High-Latitude Electric Field into Low Latitudes

One of the important problems associated with magnetospheric and ionospheric convection concerns how and under what circumstances electric fields of high-latitude origin can, or cannot, penetrate deep into lower latitudes (e.g. Swift 1971; Pellat and Laval 1972; Jaggi and Wolf 1973; Volland, 1973; Maltsev 1974; Wolf 1974; Behnke et al. 1985).

5.3.1 Substorm Effects

The earlier observational study of low-latitude effects of substorm-associated electric fields by Carpenter and Kirchhoff (1975) showed, by comparing the electric fields observed by two incoherent scatter radars at Chatanika and Millstone Hill, that the daily variations at the two radar sites are quite similar. Kirchhoff and Carpenter (1976) examined the daily variations in ionospheric drift velocities at Millstone Hill and found that daily variations of the electric field during high geomagnetic activity followed basically the usual convection pattern but included large day-to-day variations. Carpenter and Akasofu (1972) showed that the westward electric field in the plasmasphere inferred from radial motions of whistler ducts is considerably intensified during substorms. Testud et al. (1975) also reported an enhancement of the westward field observed by an incoherent scatter radar at Saint Santin, France, but suggested that not all substorms cause the enhancement.

In addition to electric field information deduced from data of the Arecibo, Puerto Rico radar, Harper (1977a,b) used variations in ground magnetic perturbations at a nearby observatory, San Juan, to discuss the origins of the ionospheric electric fields and currents at mid-latitudes during both quiet and disturbed periods. It was found that the variations in the ground magnetic field on the very disturbed days did not appear to be primarily due to overhead ionospheric currents. Using radar measurements at Saint Santin, Blanc et al. (1977, 1980) and Blanc (1978) reached the conclusion that while low-latitude extension of the high-latitude convection electric field is not always associated with substorms, it occurs in conjunction with the development of the partial ring current, implying that large dawn-dusk asymmetry in ionosphere–magnetosphere coupling during substorms may be responsible for the lack of electric field shielding at mid-latitudes. Furthermore, Smiddy et al. (1977) conducted a DC electric field experiment on a polar-orbiting satellite (S3-2). Intense, localized electric fields directed poleward in the premidnight sector near the ionospheric projection of the plasmapause were shown to be related to substorm activity. It is possible, however, to find many substorms which have no apparent effects on low-latitude electric fields, suggesting that some complicated processes may regulate the efficiency of the direct electrical coupling between high and low latitudes (Fejer et al. 1979; Gonzales et al. 1983; Mazaudier 1985; Fejer 1986; Tanaka 1986).

5.3.2 Source Mechanisms

As pointed out by Blanc and Richmond (1980), there are two generalized source mechanisms which can, at least qualitatively, account for the temporal and spatial behaviour of the electric field at middle and low latitudes during geomagnetically active periods. The first mechanism is the so-called magnetospheric dynamo, in which dynamic interactions occur between the solar wind and the magnetosphere and further the high-latitude ionosphere. Parts of these currents, associated with their electric fields, penetrate directly into lower latitudes through the conducting ionosphere. This type of ionospheric, low-latitude response to the high-latitude electric fields that couple to the magnetosphere via field lines is the result of what we call "direct" penetration of the convection electric fields into lower latitudes. This is equivalent to the direct closure of auroral currents through the ionosphere or to the leak or non-shielding of electric fields in the distant plasma sheet toward lower L shells (Vasyliunas 1972; Wolf 1975). Penetration of this type can be expected during periods of rapid changes in electric fields in the magnetosphere–ionosphere system, such as DP 2 (Nishida 1968a,b) and substorms.

The second mechanism, studied extensively by Blanc and Richmond (1980), is called the ionospheric disturbance dynamo. The thermospheric wind produced by auroral heating is in charge of altering the global circulation pattern and consequently generating electric fields and currents in the middle and low latitude regions by means of ionospheric dynamo action. It is thus natural to infer that the time scales involved in the second mechanism are much longer, at least several hours, than those of the first mechanism. In the last decade it has become possible to directly measure the ionospheric electric fields and currents by means of incoherent scatter radars. This enables us to assess the relative importance of these two mechanisms quantitatively under various magnetospheric and ionospheric conditions.

Outstanding advances include the volumes of data obtained from the Saint Santin radar along with an extensive, semi-analytical treatment of magnetospheric convection, including the effects of spatial variations of ionospheric conductivities (Senior and Blanc 1984). Their calculations show that the enhancement of the auroral conductivities by electron precipitation significantly increases both the characteristic duration time and the degree of penetration of the electric fields into the mid-latitudes. The distribution of the calculated electric fields caused by a sudden increase in the dawn-dusk polar cap electric potential is in good agreement with the available statistical models of the disturbance electric fields. Figure 5.6 shows the calculated electric fields in two components, both at the initial time and after the steady state is reached at 56° and 44° latitude along with relevant observations at Millstone Hill and Saint Santin, respectively. Senior and Blanc (1984) constructed a self-consistent model of magnetospheric convection which reflects the effect of latitudinal and longitudinal variations of ionospheric conductivities. Imposing a dawn-to-dusk potential drop across the magnetotail, the motions of the ring current and the associated field-aligned currents have been computed. By combining magnetic records with electric field data from two radars at auroral and middle latitudes, Mazaudier et al. (1984)

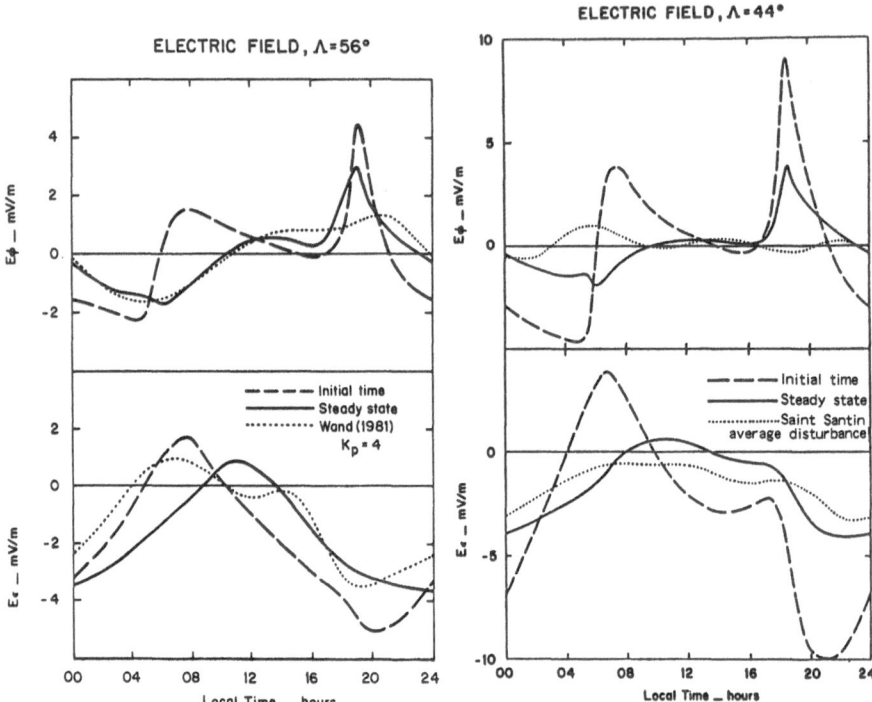

Fig. 5.6. *Left,* Initial-time and steady-state distributions of the ionospheric electric field produced at 56° invariant latitude for a 50-kV potential drop across the polar cap in the Senior and Blanc model (1984). The initial-time amplitudes are divided by a factor of two to facilitate the comparison. The *dotted curves* represent Wand's (1981) model of the disturbance electric field over Millstone Hill for $Kp = 4$. *Right,* For the latitude of Saint-Santin, 44°. The *dotted curve* represents Blanc's (1983a) model of the average disturbance E field, which was obtained by subtracting from each other the average electric field for $Kp > 3$ and for $Kp < 2^+$. (After Mazaudier et al. 1984)

showed that once the auroral conductivities can be deduced from comparison of the electric and magnetic field data at auroral latitudes, good agreement is found between the experimental latitudinal profile and the prediction of the Senior and Blanc model.

Several case studies using the Saint Santin radar data have clearly revealed the significance of either of the two mechanisms (magnetosphere–ionosphere coupling and ionosphere–thermosphere coupling) operating during disturbed periods, perhaps depending on the time scale of the events. In particular, Mazaudier (1985) pointed out that although signatures for the direct penetration of the electric fields can indeed be observed, some complexities do exist in the sense that there is no one-to-one correspondence between the high-latitude electric fields or currents and the low-latitude electric field. The complexities appear to relate to changes in the solar wind and the IMF. It is, however, also possible to identify the "delayed" response of the ionospheric dynamo to auroral electrojet activity during major magnetic storms.

5.3.3 Global Patterns of Ionospheric Fields

During the last decade, a number of simulation schemes have been developed to obtain the world pattern of electric fields and currents in the ionosphere, including the electrical coupling between high and low latitudes (e.g. Yasuhara et al. 1975; Maekawa and Maeda 1978; Nopper and Carovillano 1978). Nisbet et al. (1978) conducted a numerical calculation of the electric fields and currents in the global ionosphere produced by field-aligned currents at auroral latitudes. A two-dimensional network was constructed by using a number of rectangular grids in which the distribution of the field-aligned currents and the ionospheric conductivity was assumed to obtain the most suitable electric potential as a whole. The model has utilized the "input" field-aligned currents reported by Iijima and Potemra (1976) as well as the conductivities developed by Kirchhoff and Carpenter (1976). It was found that the electric fields due to the high-latitude, field-aligned currents do indeed penetrate down to the equator but their values at the equator are reduced by more than two orders of magnitude. Nopper and Carovillano (1978) made a similar calculation of the ionospheric electric fields and currents in relation to the field-aligned currents. It was demonstrated that changes in the field-aligned current pattern during disturbed periods can account for equatorial fluctuations in the electric fields, indicating that the magnetospheric dynamics that produce the field-aligned currents, given as the driving forces in their calculations, have a direct effect on the equatorial ionosphere through the ionospheric conductivity. However, in their calculations, the difference between quiet and disturbed conditions was modeled by changing only the location and strength of the field-aligned currents, not by changing the conductivity distribution. This assumption seems to be in contrast to recent observations that, in substorm-disturbed regions, the ionospheric conductivity is enhanced by a factor of 10 or more by auroral particle bombardment. Accordingly, it was not possible in their simulations to differentiate properly the roles of these two important factors (i.e. the field-aligned currents in the auroral belt and the enhanced ionospheric conductivity) when discussing the "polar–equatorial" coupling during active periods.

Kamide and Matsushita (1981) examined in quantitative detail the electrical coupling of high latitudes and low latitudes for selected models representing quiet and disturbed conditions. They investigated the way in which the intensity and distribution of the field-aligned currents as well as the degree of auroral enhancement affect the rate of latitudinal decrease in the electric fields toward low latitudes. It was found that the efficiency of this penetration is controlled strongly by the relative location of the ionospheric conductivities and field-aligned currents at auroral latitudes. Although in their calculations these two quantities are taken to be independent parameters in the system under study, there is evidence that significant portions of the field-aligned currents are actually carried by the precipitating electrons themselves which, in turn, ionize the ionospheric particles, resulting in conductivity enhancements.

Most of these calculations were made only for the thin-shell ionosphere by assuming the distribution of the field-aligned currents without discussing their origin in the magnetosphere. Since the ionosphere and magnetosphere are electrically linked through the field-aligned currents, the penetration of the electric field of high-latitude origin into lower latitudes is equivalent to the penetration of magnetospheric convection into lower L shells. It is therefore important to examine ionospheric and magnetospheric processes together.

5.3.4 Shielding of Convection Fields in the Magnetosphere

There has been much interest in determining the extent to which the inner magnetosphere is shielded from the convection electric field in the magnetosphere (e.g. Kivelson 1976). Some observations indicate that at quiet times the region inside the plasmapause is rather thoroughly shielded by the convection field, suggesting that the ionosphere at middle and low latitudes is dominated by dynamo electric fields associated with ionospheric winds (Matsushita and Mozer 1973; Carpenter and Seeley 1976; Richmond 1976; Harper 1977a,b). However, during intense storms and substorms, the effects of the shielding tend to diminish, bringing the convection electric field closer to the Earth than during quiet times (e.g. Maynard and Grebowsky 1977).

Following the suggestion by Volland (1973) and Stern (1975), the electrostatic potential ϕ in the equatorial plane of the magnetosphere can be analytically represented by

$$\phi = CL^{\gamma} \sin \lambda,$$

where C is a constant which is proportional to the cross-tail potential drop, L is the L shell parameter in Earth radii, and γ specifies the shielding effect of the electric field toward low L distances ($\gamma = 1$ for a uniform electric field). Although previous convection studies assumed that the convection electric field is uniform in space (e.g. Chen 1970), more recent studies have suggested that except for very quiet periods, $\gamma = 2$ may more adequately represent the observed L dependence on the electric field in the vicinity of the plasmapause (Heppner 1972b; Volland 1973; Maynard and Chen 1975; Stern 1975; Cowley and Ashour-Abdalla 1976a,b; Ejiri et al. 1978; Baumjohann et al. 1985).

Kaye and Kivelson (1979, 1981) tried to obtain the shielding factor as a function of the Kp index. In particular, Kaye and Kivelson (1981) used electric probe measurements aboard the OGO 6 spacecraft in the dawn-dusk meridian at altitudes between 400 and 1100 km. On the other hand, Ejiri (1978) used plasma data at the plasmapause crossing in the evening sector by Explorer 45 in the equatorial plane to predict the shape of the L-dependence on the electric field in the magnetosphere. It was found statistically by both groups that the shielding factor γ decreases, on average, with increasing Kp. However, we need to note the following two points which must not be ignored when discussing the electric field configuration under disturbed conditions, but which are not properly taken into consideration as long as the 3-h index Kp is used: (1) there must be different time constants, depending on different local times during which the electric field near the Alfvén layer reaches a steady state (Jaggi and Wolf 1973); (2) substorms are essentially time-dependent phenomena, indicating that the enhancement of the electric field in the distant tail takes place suddenly in conjunction with onset of the substorm—this, by no means, signifies that throughout the Kp interval (3 h) the substorm field is nearly constant.

The simulation model called the RCM (see Sect. 4.2) was applied to a substorm event (Harel et al. 1981b). It was found that, although the low-latitude ionosphere is shielded effectively from the high-latitude electric field during the event, the greatest leakage through the shielding (i.e. the breakdown of the shielding) tended to occur as a result of the rapid conductivity changes at onset of the substorm. The inner edge of the plasma sheet, which generally causes the field shielding, apparently does not have time to react to these changes.

Finally, we note that there are at least two factors which independently control the penetration to lower latitudes of an enhanced electric field imposed at high latitudes during substorms: one is an overall increase in the net field-aligned current flowing in the auroral latitudes, and the other is the field-aligned currents near the equatorward edge of the auroral belt, which usually have a lower intensity than those in the poleward half. These variable quantities are related to magnetospheric parameters and are manifested, respectively, in terms of an enhanced cross-tail electric field associated with the southward interplanetary magnetic field and the efficiency of the charge escape (or field-aligned currents) flowing from the Alfvén layer to the ionosphere. The former produces the large electric field at middle and low latitudes as well as at high latitudes. However, the fact that large electric fields are observed at middle and low latitudes is insufficient evidence for the penetration efficiency of electric fields of high-latitude origin. Such penetration can only be confirmed by a careful examination of the latitudinal variation. Since only the net field effect, resulting from complicated processes in the ionosphere and magnetosphere, can be observed at any one location at middle and low latitudes, it is at present difficult to distinguish between these two causes in actual electric field data.

5.4 Relative Importance of Conductivities and Electric Fields

5.4.1 Simultaneous Measurements of Ionospheric Parameters

Only during the last decade have several new powerful techniques become available for studying the three-dimensional current system and, as a result, our knowledge of the large-scale current distribution in the ionosphere and magnetosphere as well as the driving electric field has greatly evolved (see Kamide 1982; Troshichev 1982), for reviews. The first-approximation model of the three-dimensional current system appears to be consistent with computer simulations and theoretical calculations (e.g. Boström 1964; Fukushima 1971; Swift 1971; Wolf 1974; Kamide and Matsushita 1979a,b; Harel et al. 1981a,b).

Crucial questions remain, however, regarding the dominant physical processes responsible for the generation of the auroral electrojets. For example, are changes in the conductivity (σ) or electric field (E) more important in producing enhanced current (J) in different regions of the ionosphere? Of course, the conductivity and electric field are not totally independent in the ionosphere, in the sense that the spatial non-uniformity of σ results in changes in E. However, our interest here is to learn the relative role of the electric field and the conductivity in intensifying the auroral electrojets. For this purpose, it is desirable to measure more than one ionospheric parameter simultaneously at more than one site (see de la Beaujardiere et al. 1977). One of the merits of the operation of incoherent scatter radars is that in certain modes of operation the radar beam is able to probe the altitude/latitude distribution of electron density and line of sight plasma drifts, from which the electric fields, conductivities, and currents in the ionosphere can be determined (see Brekke et al. 1974; Banks and Doupnik 1975; Vondrak and Rich 1982; Vondrak 1983; Robinson et al. 1985a,b; Senior et al. 1987). Using this advantage, the latitudinal structure of the electric fields and conductivities was examined by Wedde et al. (1977), Horwitz et al. (1978), and Vickrey et al. (1981), although these studies put their main emphasis on the latitudinal distribution on an event-by-event basis.

In this section, we present synoptic Chatanika radar observations of the latitudinal structure of the auroral electrojets in a wide local time range. In this way, we can assess the relative importance of the electric fields and conductances at different auroral latitude locations. Kamide and Vickrey (1983) selected eight sets of data from continuous observations of more than 12 h in duration. For most of the 8 days, data from the IMS Alaska meridian chain of magnetometers were also available.

To show that the magnitude of the auroral electrojets is dominated by different ionospheric parameters in different regions, Kamide and Vickrey (1983) examined the local time dependence of these parameters by using a bulk data set representing a variety of substorm activities and local time sectors. For each of the cases, the location of the electrojet center (eastward or westward) was first identified. Then, the Hall conductivity and the north-south electric field were determined along with the geomagnetic local time (MLT) at the electrojet center. Figures 5.7 and 5.8 show the statistical results of this study.

Figure 5.7a is a scatter plot of the Hall conductivity and the east-west ionospheric current density at the center latitude of the auroral electrojets. From this relationship, it is possible to gain some insight into the degree of the conductivity contribution to the auroral electrojet. Different symbols are used to distinguish the eastward and westward electrojets. Further, the points for the westward electrojet are grouped into two categories corresponding to times before and after 0300 MLT. The choice of 0300 MLT as a dividing time is somewhat arbitrary but turned out to best order the points as a whole.

In Fig. 5.7b the north-south electric field is plotted against the east-west ionospheric current in a format similar to Fig. 5.7a. From this, we can discern the different behaviour of the electric field in the different temporal regimes of the auroral electrojets. In spite of the considerable scatter in the plots, it is apparent that the eastward electrojet in the evening sector and westward electrojet during local times after 0300 MLT have statistically a common character: namely, when the current intensity is relatively small, the increase in the current density appears to be caused by an increase in both the conductivity and the electric field. It is reasonable that $\Sigma_H > \Sigma_P$, but Σ_P is certainly not negligible. We also note that Σ_y (north-south electric field) remains at the level of 10–30 mV/m, even when J_x (east-west current) approaches zero.

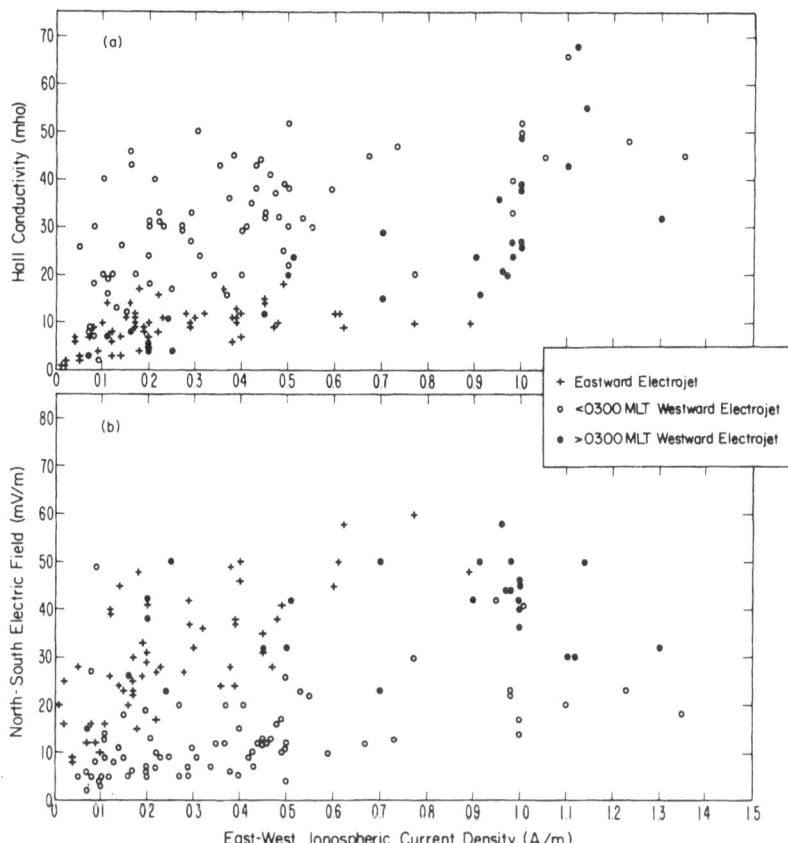

Fig. 5.7. a Dependence of the east-west ionospheric current on the Hall conductivity. **b** Dependence of the east-west ionospheric current on the north-south electric field. *Different symbols* are used to differentiate the eastward and westward electrojets. Furthermore, the westward electrojet is grouped into two jets, depending on whether it occurred before or after 0300 MLT. All points represent quantities at the latitudinal center of the auroral electrojets. (Kamide and Vickrey 1983)

It should be pointed out that between the eastward electrojet and the late morning (> 0300 MLT) westward electrojet, there are some differences as well. First, in Fig. 5.7a the strength of the eastward electrojet does not become very large, say, more than 1 A/m. Second, the conductivity in regions where the eastward electrojet exceeds 0.2 A/m does not increase very much with increasing current strength. This means that the increase in the current density is accomplished mainly by the increase in the northward electric field. On the other hand, in the region of the westward electrojet in the late morning sector, the conductivity as well as the southward electric field still continues to increase substantially with increasing current strength.

However, such differences are relatively minor, when contrasted with differences between the eastward and westward electrojets in the midnight sector

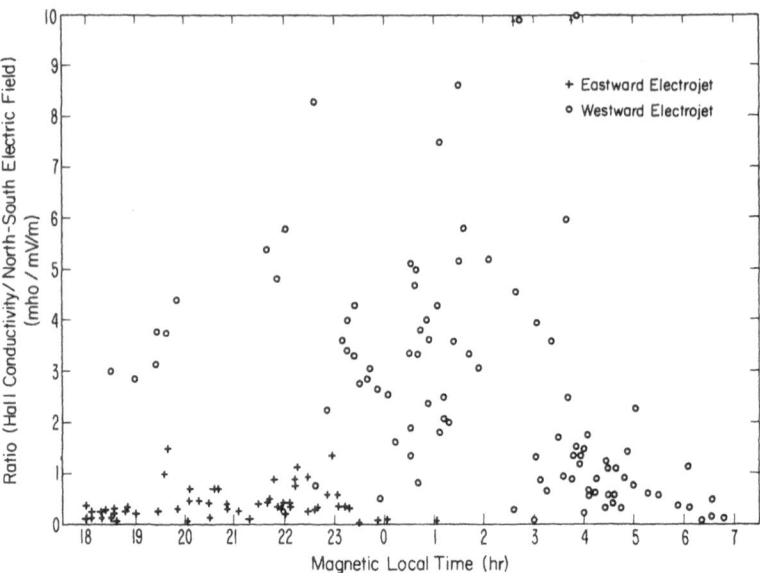

Fig. 5.8. Ratio of the Hall conductivity (in S) to the strength of the north-south electric field (in mV/m) in the eastward and westward electrojets, plotted as functions of magnetic local time

earlier than 0300 MLT. It is a striking feature that the points corresponding to the westward electrojet in the premidnight and early morning sectors are well separated from those for the eastward electrojet. Conductivity values for the midnight sector westward electrojet are significantly higher, on the average, than those for the same magnitude of the eastward electrojet and for the westward electrojet in the late morning sector. This feature is consistent with the observations of Brekke et al. (1974), Horwitz et al. (1978), and Vickrey et al. (1981). For example, in an evening eastward or late morning westward current of 0.5 A/m, a Hall conductivity of 10–20 S is typical, while the Hall conductivity ranges from 20 to 50 S for the same magnitude of the midnight sector westward electrojet. In contrast, the intensity of the southward field in the midnight sector westward electrojet is relatively small ($\sim 10\,\text{mV/m}$).

In order to clarify further the different roles and the relative importance of the conductivity and electric field in the different electrojet regions, we plot in Fig. 5.8 the ratio $\Sigma_H/|E_y|$ as a function of magnetic local time. If we use commonly adopted units, S and mV/m, the average ratio for the eastward electrojet is found to be 0.40, whereas that for the midnight sector westward electrojet before 0300 MLT is 3.71. The Hall conductivity of the midnight sector westward electrojet is sometimes 15 times larger than that in the eastward electrojet. The average ratio of Σ_H and E_y for the westward electrojet in the late morning sector ($= 1.39$) is smaller than the midnight electrojet but somewhat larger than that in the eastward electrojet.

It is clear that in the region of the eastward electrojet in the evening sector, the northward electric field is the main contributor to the magnitude of electrojet current, in the sense that the field magnitude is larger compared to the southward field magnitude in the westward electrojet. It is thus possible that the intensification of an already moderate eastward electrojet is mainly caused by an enhancement of the northward electric field. The eastward electrojet is known to develop in the diffuse aurora (Tsunoda et al. 1976a,b) that is probably caused by relatively low-energy precipitating particles. It is important to note that in the region of the eastward electrojet, the Hall and Pedersen conductivities seem to change in unison (Vickrey et al. 1981), but the Pedersen currents are associated with the northward electric field connected to field-aligned currents in this region (Baumjohann et al. 1980).

5.4.2 Two Electrojet Modes

A point of interest is that there are essentially two modes of the westward electrojet; one in which the contributions to the electrojet magnitude are "conductivity dominant" and the other "electric field dominant" (as discussed above for the eastward electrojet). The exact classification into these two modes of the westward electrojet using observed data is difficult, however, because the corresponding currents are contiguous everywhere. The westward electrojet near midnight and in early morning hours is characterized mainly by the relatively high Hall conductivity, whereas the westward electrojet in the late morning sector is dominated by the large southward electric field.

The latter behaviour is similar to that of the eastward electrojet in the evening sector, although the sense of the electric field is of course reversed. However, an important difference is that the Hall conductivity can become high in the late morning westward electroject (as high as 50 S), while the maximum conductivity for the evening eastward electrojet is, in general, less than 20 S. The high conductivity values in the morning electrojet are probably generated by keV electron precipitation in patchy auroras, moving eastward from the midnight sector with the development of substorms. In other words, both the southward field and Hall conductivity appear to be important contributors to the intense westward electrojet in the late morning sector.

5.4.3 Latitudinal Cross-Sections of the Auroral Electrojets

It is also important to point out that through a recent series of improvements of the magnetogram-inversion techniques (see Sect. 4.2), a number of important characteristics, which were not explicit in earlier studies, have become evident, even though the global patterns in individual cases are quite variable depending upon solar wind conditions and geomagnetic activity. One of the characteristics seen repeatedly in the deduced diagrams of the global electrodynamic features is that a significant amount of ionospheric currents can flow in regions where auroral activity and the corresponding conductivity are quite low. That is, the region of intense auroral electrojets is not completely the same as the region of high conductivity and that of the large electric field, since there are latitudinal shifts among these three quantities. This effect is particularly significant over the poleward half of the westward electrojet in the morning sector, where the electric field combined with relatively low conductance is the main contributor to the ionospheric current. On the other hand, the equatorward half of the auroral electrojet in the morning sector is dominated by conductivity enhancements, signalling high auroral activity in this latitudinal regime. Such findings are hardly expected from studies in which statistically derived models of the ionospheric conductivity and of the electric field are independently given.

Figure 5.9a shows latitudinal profiles of the three quantities (the electric field, the conductance, and their product, the current) overlaid for 0400 MLT (magnetic local time). These quantities have been deduced from the best-optimized result using nearly simultaneous data of DE-1 auroral imagery and of DE-2 ion drift, in addition to ground magnetometer records for the maximum epoch of an intense substorm (Kamide et al. 1989). It is clear that the electric field peaks poleward of 69° latitude, whereas the highest conductivity occurs somewhat

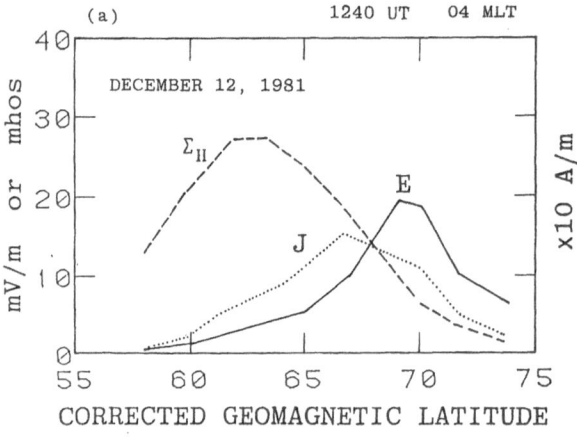

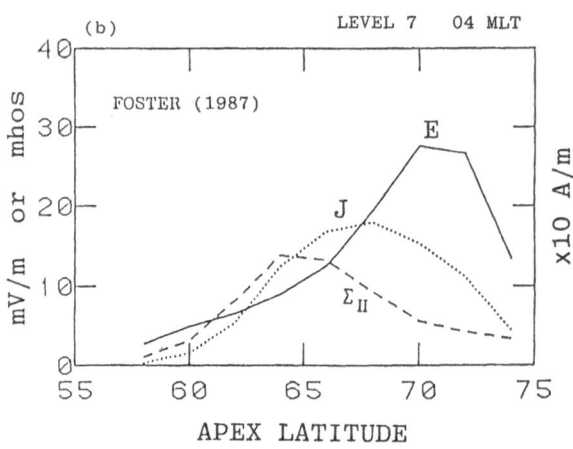

Fig. 5.9. a Individual latitude distribution of the electric field E, the Hall conductivity Σ_H, and the Hall current J for 0400 MLT, calculated through the magnetogram-inversion technique for 1252 UT, 12 December 1981. Vertical scale units: E, mV/m; Σ_H, S; J, 0.1 A/m. **b** Statistical profile of the three quantities (after Foster 1987). Foster et al. (1986) obtained the distribution of ionospheric convection determined from Millstone Hill radar data for different intensities of auroral particle precipitation, using a particle precipitation index based on spacecraft observations. This diagram is for index level 7 (see Fuller-Rowell and Evans (1987) for details of the index)

southward of it, extending to 64° latitude in this particular example, implying that enhancements in the electric field and in auroral activity control the strength of the electrojet current separately. The peak of the westward electrojet is located between the two peaks.

Figure 5.9b is a representative sample of the latitudinal profiles of the three parameters (Foster 1987) for 0400 MLT for active times, which are based on a statistical study of the convection electric fields from the Millstone Hill radar and particle precipitation from the NOAA/TIROS satellites (Foster et al. 1986). Although the exact locations of the three quantities and their magnitudes are different in Fig. 5.9a and b, the individual and statistical profiles, respectively, the separation of the electric field and conductivity peaks by several degrees is consistently noted. This offset is important because it has been a common practice to assume tacitly that the *auroral* electrojets, as the terminology implies, are flowing in the region of high conductance where the auroral luminosity is highest.

It is noted that an earlier statistical study of the auroral electrojets at different local times indicated that the auroral electrojets may have two different elements: the electric field-dominant electrojet and the conductivity-dominant electrojet (Kamide and Vickrey 1983). Combining this statistical indication and Fig. 5.9, it is inferred that the auroral electrojets do not grow and decay as a whole, but only the portion in the midnight sector and the equatorward half of the currents is enhanced at the time of auroral breakup associated with the onset of the expansion of substorms while the other portion is controlled by the electric field, which does not primarily reflect auroral activity.

Illustrating the different roles of ionospheric conductivities and electric fields in controlling the development of the auroral electrojets in different local-time sectors, Fig. 5.10 is a schematic diagram showing that the auroral electrojets can have two distinct mechanisms, one in which the contributions to the electrojet magnitude are "conductivity-dominant" and the other, "electric field-dominant". The strict differentiation into these two modes of the electrojet using observed data is impossible, because these currents are continuously connected with each other in a very complicated manner everywhere. The westward electrojet near midnight and in the early morning hours is characterized mainly by the relatively high Hall conductivity, whereas the westward electrojet in the late morning sector is dominated by the large southward electric field. The latter behaviour is similar to that of the eastward electrojet in the evening sector. This leads us to speculate that the eastward electrojet in the evening sector and the late morning portion of the westward electrojet constitute a pair of currents generated through one mechanism relating to the global convection electric field.

The Harang discontinuity is clearly manifested as a switch from the eastward to the westward electrojet, and in more detail as a switch from the "electric-field-dominant" to the "conductivity-dominant" electrojet. It is the region where auroral breakup occurs. In reality, the boundary between the conductivity-dominant and the electric field-dominant westward electrojets in the morning sector is not as clear, as indicated in Fig. 5.10. However, an important point is that, although it has been a common practice to assume that the westward electrojet is associated during substorms with a conductivity enhancement, part

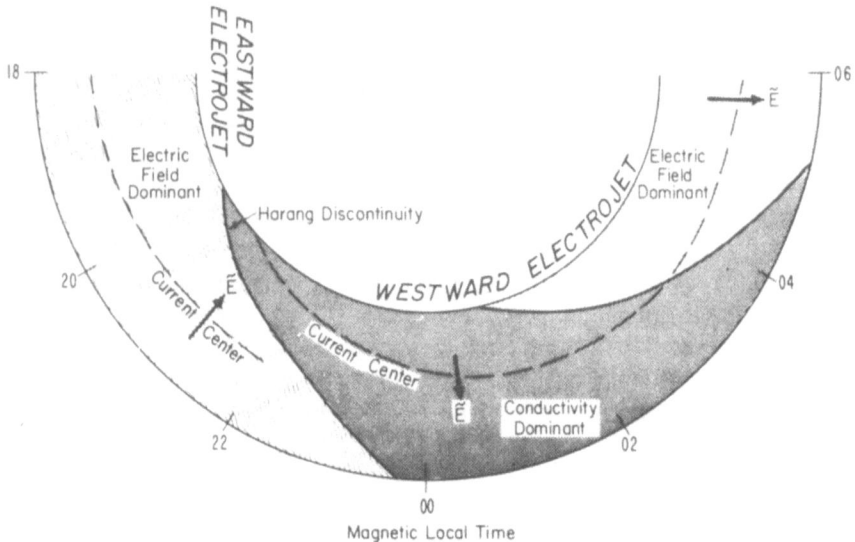

Fig. 5.10. Schematic diagram of the auroral electrojets showing the different roles of ionospheric electric fields and conductivities in the auroral electrojets at different latitudes and local times. The Harang discontinuity separates the eastward electrojet to the west, and the westward electrojet to the east. This diagram illustrates that there are two components within the westward electrojet: one is conductivity-dominant in the midnight to early morning sectors, and the other is electric field-dominant primarily in the late morning sector

of the westward electrojet in the later morning sector can be intensified without having high conductivity values. In particular, the latitudinal profile of the westward electrojet in the early morning sector indicates that its poleward portion has a relatively strong southward electric field, while its equatorward portion has a relatively high conductivity (Senior et al. 1982; Kamide and Vickrey 1983).

The implications of these two elements are interesting if they are viewed as two physical processes for solar wind–magnetosphere energy coupling, i.e. the directly driven and loading-unloading processes (Akasofu 1981; Rostoker et al. 1987a). That is, the "electric field" electrojet may represent directly the effects of the solar wind-magnetosphere dynamo, whereas the "conductivity-rich" electrojet may be a manifestation of plasma instability processes internal to the magnetosphere, relating to intense and sporadic auroral activity during the expansion phase of substorms. This will be discussed below.

5.4.4 Implications for Substorm Dynamics

More recent studies (e.g. Ahn et al. 1984, 1989; Kamide and Baumjohann 1985) have dealt with the characteristic features of substorms by using a number of individual snapshots of the large-scale distribution of high-latitude electric fields and currents at the key times and epochs of the substorms. It has been noted that these global patterns change quite dynamically in association with substorm times, or with the substorm phases. Thus, to understand the energy flow from the solar wind through the magnetosphere to the ionosphere on an individual basis, it is essential to describe in detail the temporal variations of the substorm current system before and during the substorms. This problem relates closely to the major question of the magnetospheric substorm as to the evaluation of the relative importance of two energy dissipation mechanisms: the "loading-unloading" and "directly driven" processes (Akasofu 1981), both of which have been proposed as playing an essential role in magnetospheric substorms.

One of the controversial arguments, in an attempt to construct phenomeno-logical substorm models in the past, results from an ambiguity where one may claim that any conclusions reached by examining a particular set of observations may not be true for every substorm. Perhaps, the simplest way to classify substorm activity is to divide it into two categories: isolated substorms and continuous activity. This distinction between the two categories would not be necessary if it were true that every continuous activity of substorms without the intervening period of calm could be explained by superposing linearly the isolated substorms (see Akasofu and Kan 1973). However, it is unlikely in continuous substorm activity that the "second" substorm, for example, takes place entirely independently of the first substorm, because at the onset of expansion of the second substorm, for example, the enhanced ionospheric conductivity and the expanded auroral oval probably still remain (Akasofu and Kan 1982). By the same token, the "third" substorm cannot be thought of without the existence of the first and second substorms, unless all these substorms are well separated for more than several substorm lifetimes. These "linear" and "non-linear" viewpoints cannot be com-pared physically in terms of only ground magnetometer signatures; thus, they await further detailed studies of the corresponding magnetospheric phenomena.

One way of looking into this important problem regarding the substorm generation mechanisms is to examine the substorm current system. It has already been shown that there are basically two types of current systems: DP 1 and DP 2 (see, for example, Troshichev et al. 1974; Baumjohann 1983; Clauer and Kamide 1985; Kamide and Baumjohann, 1985) indicating that these two modes can coexist at all times during disturbed periods and the relative strength of these currents varies from time to time, making individual current patterns very complex. Furthermore, the pattern of "enhanced" DP 2 has been shown to appear at times throughout substorm activity, although the DP 2 current system was originally distinguished from the DP 1 system on the basis of whether or not it is accompanied by a current concentration at auroral latitudes caused presumably by auroral enhancements (see Obayashi and Nishida 1968). Kamide and Baumjohann (1985) showed that such enhanced DP 2 patterns prevail during the recovery phase of isolated substorms and also throughout continuous sub-storm activity.

Baumjohann (1983) summarized evidence that during periods of enhanced magnetospheric activity the ionospheric current flow is affected in two ways. First, due to direct energy input from the solar wind into the magnetosphere, and thus enhanced magnetospheric convection, the current flow in the auroral electrojets increases. This may be seen as the DP 2 current system. As shown in the left panel of Fig. 2.6, this system is dominated by the eastward electrojet in the evening sector and the westward electrojet in the late morning sector, both of which are controlled primarily by the electric field. Second, the sporadic release of energy previously stored in the magnetotail leads to the formation of the substorm current wedge with strongly enhanced westward current flow in the region of active breakup auroras and the westward traveling surge around midnight, i.e. it leads to the formation of the DP 1 current system (Fig. 2.6, right panel). This supports the idea that the conventional DP 2 currents are governed by electric field enhancements, while the substorm DP 1 currents are mainly caused by strong conductivity increases accompanying the breakup auroras. The conductivity enhancement must be closely related to the disruption of the tail current and the subsequent formation of the so-called three-dimensional current wedge.

If we associate the DP 2 currents with the above-mentioned directly driven substorm process and the DP 1 currents with the unloading process, one can state that especially on days with continuous activity and nearly persistent energy input, both processes are typically operating at the same time. The relative importance of the two processes varies significantly from time to time. It is conceivable that this is the reason why substorm signatures, such as the distribution of magnetic perturbation vectors and electric fields and its time changes, are quite complicated. The term "unloading" pertains to the deposition of stored energy into the auroral ionosphere as well as into the ring current. Although it has long been argued that certain substorm-associated phenomena are consistent or inconsistent with either the directly driven process or the loading–unloading process, one could resolve such controversial argument by understanding that the magnetospheric substorm involves the two processes simultaneously.

An important point, however, is that during some interval associated with the isolated substorm, nearly pure DP 2 and DP 1 systems can be observed, as demonstrated by Nishida and Kamide (1983), and by Clauer and Kamide (1985). Such intervals are particularly interesting since certain physical processes associated with each pattern can be isolated without being contaminated by the other part. A typical DP 2 pattern is often seen during the period preceding the onset of major substorm expansion (Pellinen et al. 1982). During such an interval, the IMF remains southward, enhancing magnetospheric convection, consistent with the dawn-dusk electric field in the polar cap. It is therefore plausible to assume that the increase in the ionospheric currents and the associated energy dissipation before the major substorm expansion is directly driven by the solar wind, in agreement with Baumjohann et al. (1981) and Nishida and Kamide (1983). On the other hand, the expansion phase can occur in conjunction with the weakening or diminishing of the DP 2 component, making the DP 1 pattern dominant. Since this expansion can sometimes occur after the IMF has turned northward (e.g. Rostoker et al. 1982; Rostoker 1983), i.e. after switch-off of the solar wind energy entry into the magnetosphere, the expansion appears to be connected to the unloading of the tail energy into the polar ionosphere (McPherron 1974).

To illustrate temporal changes in the two basic processes (the directly driven and loading–unloading processes, or DP 2 and DP 1 respectively), Fig. 5.11

shows a schematic diagram of AU/AL variations for a canonical, isolated substorm with special emphasis on the dominant physical cause occurring presumably in the magnetosphere–ionosphere system. The corresponding correlation with interplanetary medium parameters is also shown. A similar representation of various ground-based magnetogram signatures was given by Rostoker et al. (1980, see their Fig. 1). It is important to note that the westward electrojet has two basic components, the relative predominance of which marks the start and the end of the substorm phases. In particular, the sudden increase in the DP 1 system centered in the midnight sector signals the onset of the expansion phase, whereas its peak corresponds to the maximum epoch of the substorm. On the other hand, the DP 2 system appears to increase and decrease slowly during the canonical, isolated substorm, making a kind of "inflection point" in the AL curve. It may be that this point marks the start of the recovery phase of the substorm event. Combining this temporal change in the eastward and westward electrojets with the spatial distribution of these currents shown in Figs. 5.11 and 2.6, one may infer that the magnetosphere–ionosphere system almost always experiences the growth and decay of the directly driven process (or of the DP 2 twin-vortex electrojet system) and it can intermittently have a burst of energy release (i.e.

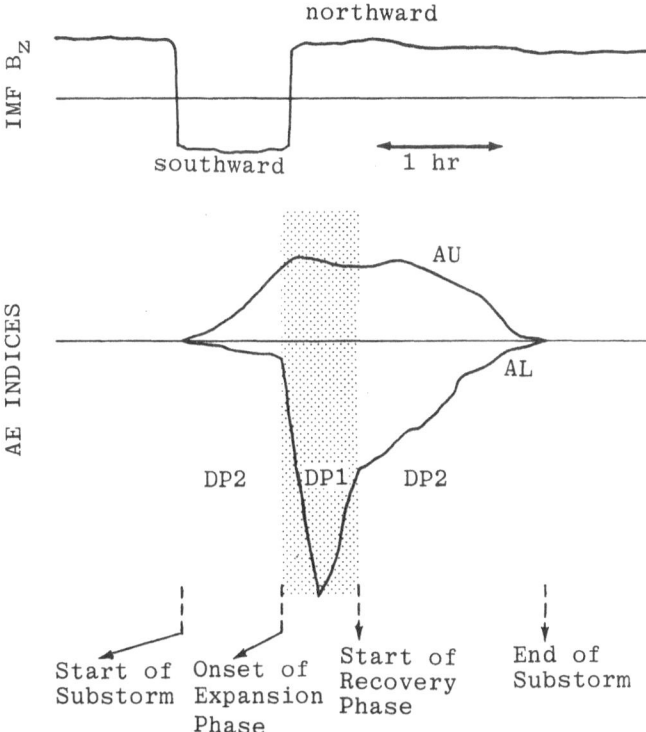

Fig. 5.11. Schematic diagram showing the development of an isolated substorm in terms of the response of the substorm phases to changes in the interplanetary magnetic field

unloading of stored energy) into the polar ionosphere, which is the onset of substorm expansion. The substorm expansion and the corresponding DP 1 enhancement are thus the signatures of sudden enhancements of the westward electrojet in the midnight sector, caused primarily by sudden conductivity enhancements associated with auroral particle precipitation (see Troshichev et al. 1974).

It should also be emphasized that, in the past, one tended to drastically simplify substorm models to discuss the basic mechanisms involved. One example is the proposed three-dimensional current system for the substorm in which the westward electrojet is connected to a downward and an upward field-aligned current flow at its eastern and western edges, respectively. This itself is not a wrong approach, because one must start with the simplest assumptions applying to the simplest examples of observations to investigate the most important and fundamental cause-and-effect parameters. However, in order to unveil what governs that current system, what particles carry the field-aligned currents, how these particles are accelerated, and what the role is of these particles in creating the enhanced ionospheric conductance, it is essential to examine the space-time distribution not only of the current, but also of some key parameters describing ionospheric quantities. For example, it is very conceivable that "high conductivity and small electric field" or "low conductivity and large electric field" may carry the same amount of electrojet currents. In these two extreme cases, we may have to invoke two very different physical processes. Some radar and rocket observations, which are capable of sensing multiple ionospheric quantities simultaneously, have investigated this problem (e.g. Senior et al. 1982; Kamide et al. 1983; Ziesolleck et al. 1983). Although their spatial scope and the number of observations are somewhat limited, it has already been shown that the relative importance of ionospheric conductivities and electric fields varies considerably, depending upon different locations within auroral electrojets.

5.4.5 Future Problems

To show the complicated nature of the interaction of electric fields, plasmas, and currents in the magnetosphere and the ionosphere via geomagnetic field lines, Fig. 5.12 summarizes some of the most important quantities in the system responsible for various physical processes. This block diagram consists of nine major boxes. The top three are essentially the magnetospheric quantities and the bottom three are the ionospheric quantities. The three boxes in between denote the space connecting the two regions, or two plasmas (hot plasma in the magnetosphere and cold ionospheric plasma), although the intermediate region at times cannot be clearly defined. The logic in this diagram is presented in such a way that, for example in the ionosphere, the electric field and the conductivity control the electric current via Ohm's law, and the electric field in the equatorial plane of the magnetosphere is mapped down to the ionosphere through geomagnetic field lines with modifications by the so-called parallel electric field.

However, the auroral ionosphere is not only a passive medium. The ionospheric electric field can be changed by polarization charges deposited by ionospheric currents near strong gradients in the conductivity distribution. The main

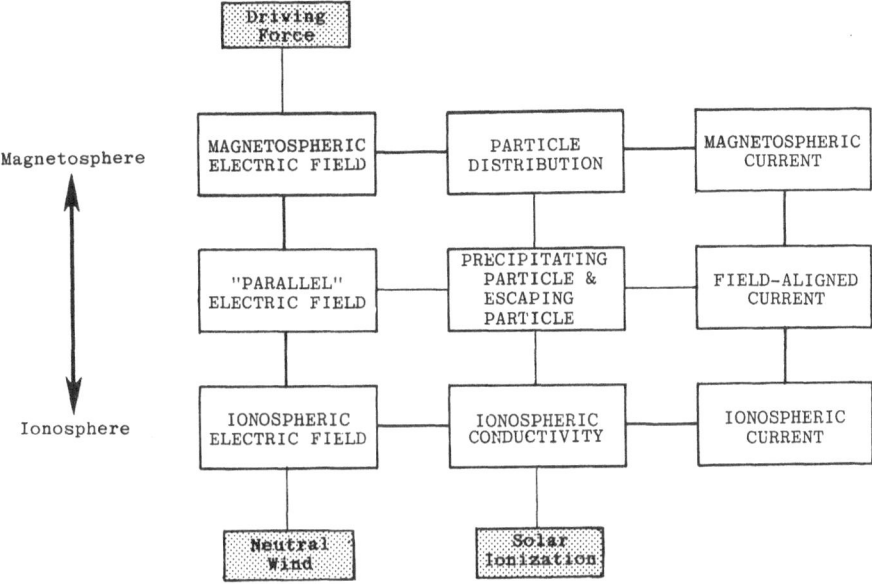

Fig. 5.12. Block diagram showing electrodynamic coupling between the magnetosphere and the ionosphere. Compare this diagram with Fig. 1.8 (Vasyliunas 1970a) and Fig. 3.1 (Harel et al. 1981a)

point of interest in Fig. 5.12 is to emphasize that, although most of the recent simulation studies treat only the outermost links of the boxes, it is essential to realize that all nine boxes are mutually coupled. For example, the rate of particle precipitation is regulated by the particle distribution in the magnetosphere, which is influenced by both magnetic and electric fields there. The precipitation is intimately related to the ionospheric conductivity. The precipitating electrons are, at the same time, the main charge carriers of field-aligned currents, which are of course regulated by the existence of the parallel electric field.

Another inevitable complication in understanding Fig. 5.12 lies in the fact that the entire system is not totally time-independent, although block diagrams of this type neglect the dynamics of the coupling processes in the magnetosphere–ionosphere system. Figure 5.12 is no exception. That is, temporal changes in these quantities that occur in the real-life magnetosphere–ionosphere system are not merely the results of repetitions of quasi-steady states separated by Δt. For instance, auroral enhancements in the ionospheric conductivity are not, in fact, directly related to and are not instantly determined by precipitating energetic electrons. As is well known, the electron density that determines the conductance is the outcome of the balance of the ionization production and recombination rates which have certain characteristic time constants. Thus, without knowing the details of the physical interaction between ionized and neutral particles, it is not possible to treat that link self-consistently and time dependently. For similar reasons, dynamic changes in the parallel electric field in charge of accelerating auroral elecrons are crucial to the unfolding of the expansive onset of substorms, which no presently available computer simulation study can explain properly.

Table 5.2. Two basic components constituting electrodynamic processes at high latitudes

Physical process	Directly driven process	Loading–unloading process
Morphological implication	Convection enhancement	Substorm expansion
Time change	Slowly varying	Spontaneous or explosive
Auroral electrojet	SD type centered at dawn and dusk	Wedge type centered in the midnight sector
Field-aligned current	Voltage source	Current source
Correlation with solar wind parameters	High	Low
With prediction filter technique	Predictable	Not predictable

In other words, we are just beginning to understand only the average state of the magnetosphere–ionosphere system in which many time-dependent and non-linear processes are taking place. At present, we do not even know how sudden is sudden, although substorm expansion is often described in terms of a sudden increase in observed parameters, such as the auroral brightness and the electrojet intensity as well as the direction of the electric field in the midnight sector.

Regardless of the complications involved in Fig. 5.12, Table 5.2 attempts to summarize the characteristics of the two basic physical mechanisms (or components) and their corresponding processes that can be observed at high latitudes and in the magnetosphere–ionosphere system. The first component is the so-called directly driven process, the electrodynamic phenomena of which are controlled by the power of the solar wind–magnetosphere MHD dynamo. The other component is the loading–unloading process that represents energy storage-release processes internal to the magnetotail. In terms of the morphology of high-latitude phenomena, these two components can be regarded as signatures of the plasma convection and of the onset of substorm expansion, respectively. The directly driven component accompanies the well-known two-cell pattern of the large-scale electrostatic potential, the size and the shape of which depend on the orientation of the IMF and the solar wind speed.

The corresponding nature of the auroral electrojets lies in the two elements of ionospheric and field-aligned currents during disturbed periods. The first is the SD-type electrojet system and the second is the localized current system near midnight. Since the time scale of the former is relatively long, i.e. > 1 h, the use of hourly values of ground-based magnetometer data in past statistical studies might have unintentionally neglected the latter component. For the same reason, correlations of the two components with solar wind parameters (such as the electric and magnetic fields in the interplanetary medium) are probably influenced by the time scale of the variations under study. If one again uses hourly or three-hourly values of the solar wind parameters and the auroral electrojet indices, the second component, i.e. the substorm expansion component, tends to be blotted out, making the correlation high. This may be the reason why significant portions of the slowly varying component of high-latitude magnetic variations are predictable by the linear prediction filter technique, which assumes that the solar wind–magnetosphere–ionosphere coupling system is linear and time-invariant. No current sophisticated observations or theories allow us to estimate the present (not averaged from past data!) prediction filter in order to reproduce spontaneous substorm expansion phenomena.

It is also interesting to infer that field-aligned currents of the directly driven component are driven primarily by voltage generators, while those of the loading–unloading component are connected to current sources in the magnetosphere. Lysak (1985) suggested that the magnetosphere–ionosphere system should behave differently as a voltage source or as a current source, depending on the scale sizes of the electrodynamic quantities. The current–voltage relationship of the three-dimensional current system is one of the fundamental properties of magnetosphere–ionosphere coupling in need of quantitative description in the future (Stern 1983; Robinson 1984; Vickrey et al. 1986; Fujii and Iijima 1987; Weimer et al. 1987).

References

Ahn B-H, Kamide Y, Akasofu S-I (1984) Global distribution of ionospheric currents during substorms. J Geophys Res 89: 1613

Ahn B-H, Kroehl HW, Kamide Y, Gorney DJ (1989) Estimation of ionospheric electrodynamic parameters using ionospheric conductance deduced from bremsstrahlung X-ray image data: J Geophys Res 94: 2565

Akasofu S-I (1974) Discrete, continuous and diffuse auroras. Planet Space Sci 22: 1723

Akasofu S-I (1976) Recent progress in studies of DMSP auroral photographs. Space Sci Rev 19: 169

Akasofu S-I (1977) Physics of Magnetospheric Substorms. Reidel, Dordrecht

Akasofu S-I (1981) Energy coupling between the solar wind and the magnetosphere. Space Sci Rev 28: 121

Akasofu S-I, Kamide Y (1985) Meridian chains of magnetometers as a powerful remote-sensing tool in determining electromagnetic quantities in the ionosphere on a global scale. EOS 53: 465–466

Akasofu S-I, Kan JR (1973) Some new thoughts on magnetospheric substorms. Radio Sci 8: 1049

Akasofu S-I, Kan JR (1982) Importance of initial ionospheric conductivity on substorm onset. Planet Space Sci 30: 1315

Akasofu S-I, Kisabeth J, Romick GJ, Kroehl HW, Ahn B-H (1980) Day-to-day and average magnetic variations along the IMS Alaska meridian chain of observatories and modeling of a three-dimensional current system. J Geophys Res 85: 2065

Akasofu S-I, Kamide Y, Kisabeth JL (1981) Comparison of two modeling methods for three-dimensional current systems. J Geophys Res 86: 3389

Alfvén H, Fälthammar CG (1963) Cosmical Electrodynamics, 2nd edn. Oxford Univ Press, Oxford

Andre D, Baumjohann W (1982) Joint Two-dimensional observations of ground magnetic and ionospheric electric fields associated with auroral zone currents. 5. Current systems associated with eastward drifting omega bands. J Geophys 50: 194

Arnoldy RL, Lewis PB, Isaacson PO (1974) Field-aligned auroral electron fluxes. J Geophys Res 79: 4208

Arnoldy RL, Moore TE, Cahill LJ Jr (1985) Low-altitude field-aligned electrons. J Geophys Res 90: 8445

Atkinson G (1970) Auroral arcs: Result of the interaction of a dynamic magnetosphere with the ionosphere. J Geophys Res 75: 4746

Atkinson G (1984) Field-aligned currents as a diagnostic tool: Result, a renovated model of the magnetosphere. J Geophys Res 89: 217

Atkinson G, Hutchinson D (1978) Effects of the day night ionospheric conductivity gradient on polar cap convective flow. J Geophys Res 83: 725

Axford WI (1969) Magnetospheric convection. Rev Geophys Space Phys 7: 421

Axford WI, Hines CO (1961) A unifying theory of high-latitude geophysical phenomena and geomagnetic storms. Can J Phys 39: 1433

Baker DN, Akasofu S-I, Baumjohann W, Bieber JW, Fairfield DH, Hones EW, Jr, Mauk B, McPherron RL, Moore TE (1984) Substorms in the magnetosphere. In: Butler DM, Papadopoulos K (eds) Solar Terrestrial Physics—Present and Future. NASA, Washington.

Baker DN, Fritz TA, McPherron RL, Fairfield DH, Kamide Y, Baumjohann W (1985) Magnetotail energy storage and release during the CDAW 6 substorm analysis intervals. J Geophys Res 90: 1205

Banks PM, Doupnik JR (1975) A review of auroral zone electrodynamics deduced from incoherent scatter radar observations. J Atmos Terr Phys 37: 951

Banks PM, Araki T, Clauer CR, Maurice JP St Foster (1984) The interplanetary electric field, cleft currents and plasma convection in the polar caps. Planet Space Sci 32: 1551

Baumjohann W (1983) Ionospheric and field-aligned current systems in the auroral zone: A concise review. Adv Space Res 2(10): 55

Baumjohann W (1986) Some recent progress in substorm studies. J Geomagn Geoelec 38: 633

Baumjohann W (1988) The plasma sheet boundary layer and magnetospheric substorms. J Geomagn Geoelec 40: 157

Baumjohann W, Friis-Christensen E (1985) Dayside high-latitude ionospheric current systems. In: Holtet I, Egeland A (eds) The Polar Cusp. Reidel, Dordrecht, pp 223

Baumjohann W, Glaßmeier KH (1984) The transient response mechanism and Pi 2 pulsations at substorm onset—review and outlook. Planet Space Sci 32: 1361

Baumjohann W, Haerendel G (1985) Magnetospheric convection observed between 0600 and 2100 LT: Solar wind and IMF dependence. J Geophys Res 90: 6370

Baumjohann W, Haerendel G (1987) Erdmagnetismus and extraterrestrische Vorgänge. Naturwissenschaften 74: 181–187

Baumjohann W, Kamide Y (1981) Joint two-dimensional observations of ground magnetic and ionospheric electric fields associated with auroral zone currents. 2. Three-dimensional current flow in the morning sector during substorm recovery. J Geomagn Geoelec 33: 297

Baumjohann W, Kamide Y (1984) Hemispherical Joule heating and the AE indices. J Geophys Res 89: 383

Baumjohann W, Paschmann G (1987) Solar wind-magnetosphere coupling: Processes and observations. Phys Scripta T18: 61

Baumjohann W, Sulzbacher H, Potemra TA (1979) Joint magnetic observations of small-scale structures in a westward electrojet with Triad and the Scandinavian Magnetometer Array. In: Japanese IMS Committee (ed) Selected Topics. Tokyo University, Tokyo, pp 49

Baumjohann W, Untiedt J, Greenwald RA (1980) Joint two-dimensional observations of ground magnetic and ionospheric electric fields associated with auroral zone currents. 1. Three-dimensional current flows associated with a substorm-intensified eastward electrojet. J Geophys Res 85: 1963

Baumjohann W, Pellinen RJ, Opgenoorth HJ, Nielsen E (1981) Joint two-dimensional observations of ground magnetic and ionospheric electric fields associated with auroral zone currents: Current systems associated with local auroral breakups. Planet Space Sci 29: 431

Baumjohann W, Haerendel G, Melzner F (1985) Magnetospheric convection observed between 0660 and 2100 LT: Variations with Kp. J Geophys Res 90: 393

Baumjohann W, Nakamura R, Haerendel G (1986) Dayside equatorial-plane convection and IMF sector structure. J Geophys Res 91: 4577

Baumjohann W, Paschmann G, Cattell CA (1989) Average plasma properties in the central plasma sheet. J Geophys Res 94: 6597

Behm DA, Primdahl F, Zanetti LJ, Arnoldy RL, Cahill LJ Jr (1979) Ionospheric electric currents in the late evening plasma flow reversal. J Geophys Res 84: 5339

Behnke R, Kelley M, Gonzales C, Larsen M (1985) Dynamics of the Arecibo ionosphere: A case study approach. J Geophys Res 90: 4448

Brin J, Hones EW Jr (1981) Three-dimensional computer modeling of dynamic reconnection in the geomagnetic tail. J Geophys Res 86: 6802

Blanc M (1978) Mid-latitude convection electric fields and their relation to ring current development. Geophys Res Lett 5: 203

Blanc M (1983a) Magnetospheric convection effects at mid-latitudes. 1. Saint-Santin observations. J Geophys Res 88: 211

Blanc M (1983b) Magnetospheric convection effects at mid-latitudes. 3. Theoretical derivation of the disturbance pattern in the plasmasphere. J Geophys Res 88: 235

Blanc M, Richmond AD (1980) The ionospheric disturbance dynamo. J Geophys Res 85: 1669

Blanc M, Amayenc P, Bauer P, Taieb C (1977) Electric field induced drifts from the French incoherent scatter facility. J Geophys Res 82: 87

Blanc M, Thomas DP, Williams PJS (1980) On the latitude variation of magnetospheric electric fields at mid-latitudes. J Atmos Terr Phys 42: 407

Block L (1972) Potential double layers in the ionosphere, Cosmic Electrodynamics 3: 349

Block LP (1975) Double layers. In: Hultqvist B, Stenflo L (eds), Physics of the Hot Plasma in the Magnetosphere. Plenum Press, New York, pp 229

Block LP (1978) A double layer review. Astro Space Sci 55: 59

Block LP, Fälthammar C-G (1976) Mechanisms that may support magnetic-field-aligned electric fields in the magnetosphere. Ann Geophys 32: 161

Boehm MH, Mozer FS (1981) An S3-3 search for confined regions of large parallel electric fields. Geophys Res Lett 8: 607

Borovsky JD (1984) The production of ion conics by oblique double layers. J Geophys Res 89: 2251

Borovsky JE, Joyce G (1983a) The simulation of plasma double-layer structures in two dimensions. J Plasma Phys 29: 45

Borovsky JE, Joyce G (1983b) Numerically simulated two-dimensional auroral double layers, J Geophys Res 88: 3116

Boström R (1964) A model of the auroral electrojets. J Geophys Res 69: 4983

Boström R (1974) Ionosphere–magnetosphere coupling. In: McCormac BM (ed) Magnetospheric Physics. Reidel, Hingham, MA, pp 45-

Boström R (1975) Mechanisms for driving Birkeland currents. In: Hultqvist B, Stenflo L (eds) Physics of the Hot Plasma in the magnetosphere. Plenum Press, New York, pp 431

Brekke A, Doupnik JR, Banks PM (1974) Incoherent scatter measurement of E region conductivities and current in the auroral zone. J Geophys Res 79: 3773

Buchert S, Baumjohann W, Haerendel G, LaHoz C, Lühr H (1988) Magnetometer and incoherent scatter observations of an intense Ps 6 pulsation event. J Atmos Terr Phys 49: 357

Buchert S, Haerendel G, Baumjohann W (1991) A model for the electric fields, currents and conductances during a Ps 6 pulsation event. J Geophys Res 95: 3733

Bujarbarua JS, Goswami KS (1985) Weak ion acoustic double layers and solitary waves on the auroral field lines. J Geophys Res 90: 7611

Burch JL (1988) Energetic particles and currents: Results from Dynamics Explorer. Rev Geophys 26: 215

Burch JL, Gurgiolo C, Menietti JD (1990) The electron signature of parallel electric fields. Geophys Res Lett 17: 2329

Burke WJ, Kelley MC, Sagalyn RC, Smiddy M, Lai ST (1979) Polar cap electric field structure with a northward interplanetary magnetic field. Geophys Res Lett 6: 21

Burke WJ, Hardy DA, Rich FJ, Sagalyn RC, Shuman B, Smiddy M, Vancour R, Wildman PJL, Kelley MC, Doyle MA, Gussenhoven MS, Saflekos NA (1984) High latitude electrodynamics: Observations from S3-2. Space Sci Rev 37: 161

Cahill LJ Jr, Greenwald RA, Nielsen E (1978) Auroral radar and rocket double-probe observations of the electric field across the Harang discontinuity. Geophys Res Lett 5: 687

Cahill LJ Jr, Arnoldy RL, Taylor WWL (1980) Rocket observations at the northern edge of the eastward electrojet. J Geophys Res 85: 3407

Cao F, Kan JR (1990) Reflections of Alfvén waves at an open magnetopause. J Geophys Res 95: 4257

Carpenter DL Akasofu S-I (1972) Two substorm studies of relations between westward electric fields in the outer plasmasphere, auroral activity, and geomagnetic perturbations. J Geophys Res 77: 6854

Carpenter DL, Kirchhoff VW (1975) Comparison of high-latitude and mid-latitude ionospheric electric fields. J Geophys Res 80: 1810

Carpenter DL, Park CG (1973) On what ionospheric workers should know about the plasmapause–plasmasphere. Rev Geophys 11: 133

Carpenter DL, Seeley NT (1976) Cross-L plasma drifts in the outer plasmasphere: Quiet time patterns. J Geophys Res 81: 2728

Chapman S (1935) The electric current system of magnetic storms. Terr Magn 40: 349

Chapman S, Bartels J (1940) Geomagnetism, vol 1. Clarendon, Oxford

Chen AJ (1970) Penetration of low-energy protons into the magnetosphere. J Geophys Res 75: 2458

Chiu YT, Cornwall JM (1980) Electrostatic model of a quiet auroral arc. J Geophys Res 85: 543

Chiu YT, Schulz M (1978) Self-consistent particle and parallel electrostatic field distributions in the magnetosphere–ionosphere auroral region. J Geophys Res 83: 629

Chiu YT, Schulz M, Fennell JF, Kishi AM (1983) Mirror instability and the origin of morningside auroral structure. J Geophys Res 88: 4041

Clauer CR, Banks PM (1986) Relationship of the interplanetary electric field to the high-latitude ionospheric electric field and currents: observations and model simulation. J Geophys Res 9: 6959

Clauer CR, Friis-Christensen E (1988) High-latitude dayside electric fields and currents during strong northward interplanetary magnetic field: Observations and model simulation. J Geophys Res 93: 2749

Clauer CR, Kamide Y (1985) DP 1 and DP 2 current systems for the March 22, 1979, substorms. J Geophys Res 90: 1343

Cowley SWH (1982) The causes of convection in the Earth's magnetosphere: A review of developments during the IMS. Rev Geophys Space Phys 20: 531

Cowley SWH (1983) Interpretation of observed relations between solar wind characteristics and effects at ionospheric altitudes. In: Hultqvist B, Hagfors T (eds) High-latitude Space Plasma Physics. Plenum Press, New York, pp 225

Cowley SWH, Ashour-Abdalla M (1976a) Adiabatic plasma convection in a dipole field: Electron forbidden-zone effects for a simple electric field model. Planet Space Sci 24: 805

Cowley SWH, Ashour-Abdalla M (1976b) Adiabatic plasma convection in a dipole field: Proton forbidden-zone effects for a simple electric field model. Planet Space Sci 24: 821

Crooker NU, Siscoe GL (1983) Ring coupling model: Implications for substorm onsets. Geophys Res Lett 10: 761

Crooker NU, Siscoe GL, Doyle MA, Burke WJ (1984) Ring coupling model: Implications for the ground state of the magnetosphere. J Geophys Res 89: 369

de la Beaujardiere O, Vondrak R, Baron M (1977) Radar observations of electric fields and currents associated with auroral arcs. J Geophys Res 82: 5051

de la Beaujardiere O, Vondrak R, Heelis R, Hanson W, Hoffman R (1981) Auroral arc electrodynamic parameters measured by AE-C and the Chatanika radar. J Geophys Res 86: 4671

Dungey JW (1961) Interplanetary magnetic fields and the auroral zones. Phys Rev Lett 6: 47

Eastman TE, Hones EW Jr, Bame SJ, Asbridge JR (1976) The magnetospheric boundary layer: Site of plasma, momentum and energy transfer from the magnetosheath into the magnetosphere. Geophys Res Lett 3: 685

Eastman TE, Rostoker G, Frank LA, Huang CY, Mitchell DG (1988) Boundary layer dynamics in the description of magnetospheric substorms. J Geophys Res 93: 14411

Ejiri M (1978) Trajectory traces of charged particles in the magnetosphere. J Geophys Res 83: 4798

Ejiri M, Hoffman RA, Smith PH (1978) The convection electric field model for the magnetosphere based on Explorer 45 observations. J Geophys Res 83: 4811

Erickson GM, Spiro RW, Wolf RA (1991) The physics of the Harang discontinuity. J Geophys Res 96: 1633

Evans DS (1968) The observations of a near monoenergetic flux of auroral electrons. J Geophys Res 73: 2315

Evans DS (1975) Evidence for low altitude acceleration of auroral particles. In: Hultquist B, Stenflo L (eds) Physics of Hot Plasmas in the Magnetosphere. Plenum Press, New York, pp 319

Evans, D.S., Maynard NC, Trøim J, Jacobson T, Egeland A (1977) Auroral vector electric fields and particle comparisons. 2. Electrodynamics of an arc. J Geophys Res 82: 2235

Fayermark DS (1977) Reconstruction of the three-dimensional current system of the high-latitude region from ground-based geomagnetic measurements. Geomagn Aeron 17: 114 (Engl Transl)

Fejer BG (1986) Equatorial ionospheric electric fields associated with magnetospheric disturbances. In: Kamide Y, Slavin JA (eds) Solar Wind–Magnetosphere Coupling. Terrapub/Reidel, Tokyo, pp 519

Fejer BG, Gonzales CA, Farbey DT, Kelley MC, Woodman RF (1979) Equatorial electric fields during magnetically disturbed conditions. 1. The effect of the interplanetary magnetic field. J Geophys Res 84: 5797

Fejer JA (1964) Theory of the geomagnetic daily disturbance variations. J Geophys Res 69: 123

Feldstein YI, Levitin AE (1986) Solar wind control of electric fields and current in the ionosphere. J Geomagn Geoelec 38: 1143

Fennell JF (1984) IMS contributions to the understanding of auroral precipitation, transport, and particle sources. In: Roederer JG (ed) Achievements of the IMS. ESA Sci Tech Publ, Noordwijk pp 731

Fontaine D, Blanc M (1983) A theoretical approach to the morphology and dynamics of diffuse auroral zones. J Geophys Res 88: 7171

Fontaine D, Blanc M, Reinhart L, Glowinski R (1985) Numercial simulations of the magnetospheric convection including the effects of electron precipitation. J Geophys Res 90: 8343

Foster JC (1983) An empirical electric field model derived from Chatanika radar data. J Geophys Res 88: 981

Foster JC (1987) Reply to Kamide and Richmond. Geophys Res Lett 14: 160

Foster JC, Doupnik JR, Stiles GS (1981) Large scale patterns of auroral ionospheric convection observed with the Chatanika radar. J Geophys Res 86: 11357

Foster JC, Holt JM, Musgrove RG, Evans DS, (1986) Ionospheric convection associated with discretee levels of particle precipitation. Geophys Res Lett 13: 656

Frank LA, Ackerson KL (1971) Observations of charged particle precipitation into the auroral zone. J Geophys Res 76: 3612

Frank LA, Ackerson KL, Yeager DM (1977) Observation of atomic oxygen (O^+) in the earth's magnetotail. J Geophys Res 82: 129

Frank LA, Craven JD, Burch JL, Winningham JD (1982) Polar views of the earth's aurora with Dynamics Explorer. Geophys Res. Lett 9: 1001

Frank LA, Craven JD, Gurnett DA, Shawhan SD, Weimer DR, Burch JL, Winningham JD, Chappell CR, Waite JH, Heelis RA, Maynard NC, Sugiura M, Peterson WK, Shelley EG (1986) The theta aurora. J Geophys Res 91: 3177

Fridman M, Lemaire J (1980) Relationship between auroral electron fluxes and field-aligned electric potential difference. J Geophys Res 85: 664

Friis-Christensen E, Wilhjelm J (1975) Polar cap currents for different directions of the interplanetary magnetic field in the Y-Z plane. J Geophys Res 80: 1248

Friis-Christensen E, Kamide Y, Richmond AD, Matsushita S (1985) Interplanetary magnetic field control of high-latitude electric fields and currents determined from Greenland magnetometer data. J Geophys Res 90: 1325

Fujii R, Iijima T (1987) The control of the ionospheric conductivities on large-scale Birkeland current intensities under geomagnetic quiet conditions. J Geophys Res 92: 4505

Fukushima N (1969) Equivalence in ground geomagnetic effect of Chapman-Vestine's and Birkeland-Alfvén's current systems for polar magnetic storms. Rep Ionos Space Res Jpn 32: 219

Fukushima N (1971) Electric current systems for polar substorms and their magnetic effect below and above the ionosphere. Radio Sci 6: 269

Fukushima N (1976) Generalized theorem for no ground magnetic effect of vertical currents connected with Pedersen currents in the uniform-conductivity ionosphere. Rep Ionos Space Res Jpn 30: 35

Fuller-Rowell TJ, Evans DS (1987) Height-integrated Pedersen and Hall conductivity patterns inferred from the TIROS-NOAA satellite data. J Geophys Res 92: 7606

Galperin YuL, Crasnier J, Kinokov YuV, Nicolaenko LM, Sinitsin M, Sauvaud JA, Khalipov VL (1977) Diffuse auroral zone. Cosmic Res 15: 421 (Engl Transl)

Gelpi C, Singer HJ, Hughes WJ (1987) A comparison of magnetic signatures and DMSP auroral images at substorm onset: Three case studies. J Geophys Res 92: 2447

Ghielmetti AG, Johnson RG, Sharp RD, Shelley EG (1978) The latitudinal diurnal, and altitudinal distributions of upward flowing energetic ions of ionspheric origin. Geophys Res Lett 5: 59

Glaßmeier K-H (1984) On the influence of ionosphere with non-uniform conductivity distribution on hydromagnetic waves. J Geophys 54: 125

Glaßmeier K-H (1987) Ground-based observations of field-aligned currents in the auroral zone: Methods and results. Ann Geophys 5A: 115

Goertz CK (1979) Double layers and electrostatic shocks in space. Rev Geophys Space Phys 17: 418

Goertz CK, Boswell RW (1979) Magnetosphere–ionosphere coupling. J Geophys Res 84: 7239

Gonzales CA, Kelley MC, Behnke RA, Vickrey JF, Wand R, Holt J (1983) On the latitudinal variations of the ionospheric electric field during magnetic disturbances. J Geophys Res 88: 9135

Gorney DJ (1987) U.S. progress in auroral research: 1983–1986. Rev Geophys 25: 555

Gorney DJ, Chiu YT, Croley DR, Jr (1985) Trapping of ion conics by downward parallel electric fields. J Geophys Res 90: 4205

Greenspan ME (1984) Effects of oblique double layers on upgoing ion pitch angle and gyrophase. J Geophys Res 89: 2842

Greenwald RA, Weiss W, Nielsen E, Thomson NR (1978) STARE: a new radar auroral backscatter experiment in northern Scandinavia. Radio Sci 13: 1021

Gurevich AV, Krylov AL, Tsedilina EE (1976) Electric fields in the earth's magnetosphere and ionosphere Space Sci Rev 19: 59

Gussenhoven MS, Hardy DA, Heinemann N (1983) Systematics of the equatorward diffuse auroral boundary. J Geophys Res 88: 5692

Gustafsson G, Baumjohann W, Iversen I (1981) Multi-method observations and modelling of the three-dimensional currents associated with a very strong Ps6 event. J Geophys 49: 138

Haerendel G (1983) An Alfvén wave model of auroral arcs. In: Hultqvist B, Hagfors T (eds) High-latitude Space Plasma Physics. Plenum Press, New York, pp. 515

Haerendel G, Paschmann G (1982) Interaction of the solar wind with the dayside magnetosphere. In: Nishida A (ed) Magnetospheric Plasma Physics. Reidel, Dordrecht, pp. 49

Haerendel G, Rieger E, Valenzuela A, Foppl H, Stenbaek-Nielson HC, Wescott EM (1976) First observation of electrostatic acceleration of barium ions into the magnetosphere. Eur Space Agency Rep ESA-SP115

Hallinan TJ, Stenbaek-Nielsen HC, Deehr CS (1985) Enhanced aurora. J Geophys Res 90: 8461

Hardy DA, Gussenhoven MS, Holeman E (1985) A statistical model of auroral electron precipitation. J Geophys Res 90: 4229

Harel M, Wolf RA, Reiff PH, Spiro RW, Burke WJ, Rich FJ, Smiddy M (1981a) Quantitative simulation of a magnetospheric substorm. 1. Model logic and overview. J Geophys Res 86: 2217

Harel M, Wolf RA, Spiro RW, Reiff PH, Chen C-K, Burke WJ, Rich FJ, Smiddy M (1981b) Quantitative simulation of a magnetospheric substorm. 2. Comparison with observations. J Geophys Res 86: 2242

Harper RM (1977a) A comparison of ionospheric currents, magnetic variations and electric fields at Arecibo. J Geophys Res 82: 3233

Harper RM (1977b) Tidal winds in the 100- to 200-km region at Arecibo. J Geophys Res 82: 3243

Hasegawa A, Sato T (1979) Generation of field-aligned current during substorm. In: Akasofu S-I (ed) Dynamics of the magnetosphere. Reidel, Dordrecht, pp. 529

Hasegawa A, Sato T (1991) Space plasma physics. 1. Stationary Processes, Springer, Berlin Heidelberg, New York

Heelis RA, Hanson WB (1980) High-latitude ion convection in the nighttime F region. J Geophys Res 85: 1995

Heelis RA, Winningham JD, Sugiura M, Maynard NC (1984) Particle acceleration parallel and perpendicular to the magnetic field observed by DE-2, J Geophys Res 89: 3893

Heppner JP (1972a) The Harang discontinuity in auroral belt ionospheric currents. Geofys Publ 29: 105

Happner JP (1972b) Electric fields in the magnetosphere. In: Dyer ER (ed) Critical Problems of Magnetospheric Physics. National Academy of Sciences, Washington DC, pp. 107

Heppner JP (1973) High latitude electric fields and the modulations related to interplanetary magnetic field parameters. Radio Sci 8: 933

Heppner JP, Maynard NC (1987) Empirical high-latitude electric field models. J Geophys Res 92: 4467

Hill TW (1983) Solar–wind magnetosphere coupling. In: Carovillano RL, Forbes JM (eds) Solar-Terrestrial Physics. Reidel, Dordrecht, pp. 261

Hoffman RA (1988) The magnetosphere, ionosphere, and atmosphere as a system: Dynamics Explorers 5 years later. Rev Geophys 26: 209

Hoffman RA, Burch JL (1973) Electron precipitation and substorm morphology. J Geophys Res 78: 2867

Hoffman RA, Evans DS (1968) Field-aligned electron bursts at high latitudes observed by OGO 4. J Geophys Res 73: 6201

Hoffman RA, Sugiura M, Maynard NC (1985) Current carriers for the field-aligned current system. Adv Space Res 5: 109

Hones EW Jr, Asbridge JR, Bame SJ (1971) Time variations of the magnetotail plasma sheet at 18 R_E determined from concurrent observations by a pair of Vela satellites. J Geophys Res 76: 4402

Hones EW Jr, Pytte T, West WI Jr (1984a) Associations of geomagnetic activity with plasma sheet thinning and expansion: A statistical study. J Geophys Res 89: 5471

Hones EW Jr, Baker DN, Bame SJ, Feldman WC, Gosling JT, McComas DJ, Zwickl RD, Slavin JA, Smith EJ, Tsurutani BT (1984b) Structure of the magnetotail at 220 R_E and its response to geomagnetic activity. Geophys Res Lett 11: 5

Horwitz JL (1984) Residence time heating effect in auroral conic generation. Planet Space Sci 32: 1115

Horwitz JL (1987) Core plasma in the magnetosphere. Rev Geophys 25: 579

Horwitz JL, Doupnik JR, Banks PM (1978) Chatanika radar observations of the latitudinal distributions of auroral zone electric fields, conductivities, and currents. J Geophys Res 83: 1463

Huang CY, Frank LA, Eastman TE (1984) High-altitude observations of an intense inverted V event. J Geophys Res 89: 7423

Hudson MK, Lotko W, Roth I, Witt E (1983) Solitary waves and double layers on auroral field lines. J Geophys Res 88: 916

Hughes TJ, Rostoker G (1979) A comprehensive model current system for high-latitude magnetic activity. I. The steady state system. Geophys J R Astr Soc 58: 525

Hultqvist B (1971) On the production of a magnetic-field-aligned electric field by the interaction between the hot magnetospheric plasma and the cold ionsophere. Planet Space Sci 19: 749

Hultqvist B (1983) On the origin of the hot ions in the disturbed dayside magnetosphere. Planet Space Sci 31: 173

Iijima T, Potemra TA (1976) The amplitude distribution of field-aligned currents at northern high latitudes observed by TRIAD. J Geophys Res 81: 2165

Iijima T, Potemra TA (1978) Large-scale characteristics of field-aligned currents associated with substorms. J Geophys Res 83: 599

Iijima T, Potemra TA, Zanetti LJ, Bythrow PF (1984) Large-scale Birkeland currents in the dayside polar region during strongly northward IMF: A new Birkeland current system. J Geophys Res 89: 7441

Iijima T, Potemra TA, Zanetti LJ (1990) Large-scale characteristics of magnetospheric equatorial currents. J Geophys Res 95: 991

Inhester B, Baumjohann W, Greenwald RA, Nielsen E (1981) Joint two-dimensional observations of ground magnetic and ionospheric electric fields associated with auroral zone currents. 3. Auroral zone currents during the passage of a westward travelling surge. J Geophys 49: 155

Jaggi RK, Wolf RA (1973) Self-consistent calculation of the motion of a sheet of ions in the magnetosphere. J Geophys Res 78: 2852

Johnstone AD (1983) The mechanism of pulsating aurora. Ann Geophys 1: 397

Kamide Y (1978) On current continuity at the Harang discontinuity. Planet Space Sci 26: 237

Kamide Y (1982) The relationship between field-aligned currents and the auroral electrojets: A review. Space Sci Rev 31: 127

Kamide Y (1988) Recent issues in studies of magnetosphere–ionosphere coupling. J Geomagn Geoelec 40: 131

Kamide Y, Akasofu S-I (1975) The auroral electrojet and global auroral features. J Geophys Res 80: 3585

Kamide Y, Akasofu S-I (1983) Notes on the auroral electrojet indices. Rev Geophys Space Phys 21: 1647

Kamide Y, Baumjohann W (1985) Estimation of electric fields and currents from International Magnetospheric Study magnetometer data for the CDAW 6 intervals: Implications for substorm dynamics. J Geophys Res 90: 1305

Kamide Y, Fukushima N (1972) Positive geomagnetic bays in evening high latitudes and their possible connection with partial ring current. Rep Ionos Space Res Jpn 26: 79

Kamide Y, Matsushita S (1979a) Simulation studies of ionospheric electric fields and currents in relation to field-aligned currents. 1. Quiet periods. J Geophys Res 84: 4083

Kamide Y, Matsushita S (1979b) Simulation studies of ionospheric electric fields and currents in relation to field-aligned currents. 2. Substorms. J Geophys Res 84: 4099

Kamide Y, Matsushita S (1981) Penetration of high-latitude electric fields into low latitudes. J Atmos Terr Phys 43: 411

Kamide Y, Richmond AD (1982) Ionospheric conductivity dependence of electric fields and currents estimated from ground magnetic observations. J Geophys Res 87: 8331

Kamide Y, Rostokerr G (1977) The spatial relationship of field-aligned currents and auroral electrojets to the distribution of nightside auroras. J Geophys Res 82: 5589

Kamide Y, Vickrey JF (1983) Relative contribution of ionospheric conductivity and electric field to the auroral electrojets. J Geophys Res 88: 7989

Kamide Y, Richmond AD, Matsushita S (1981) Estimation of ionospheric electric fields, ionospheric currents, and field-aligned currents from ground magnetic records. J Geophys Res 86: 801

Kamide Y, Akasofu S-I, Ahn B-H, Baumjohann W, Kisabeth J (1982a) Total current of the auroral electrojet estimated from the IMS Alaska meridian chain of magnetic observatories. Planet Space Sci 30: 621

Kamide Y, Ahn B-H, Akasofu S-I, Baumjohann W, Friis-Christensen E, Kroehl HW, Maurer H, Richmond AD, Rostoker G, Spiro RW, Walker JK, Zaitzev AN (1982b) Global distribution of ionospheric and field-aligned currents during substorms as determined from six IMS meridian chains of magnetometers: Initial results. J Geophys Res 87: 8228

Kamide Y, Kroehl HW, Hausman BA, McPherron RL, Akasofu S-I, Richmond AD, Reiff PH, Matsushita S (1983) Numerical modeling of ionospheric parameters from global IMS magnetometer data for the CDAW-6 intervals. Rep UAG-88. World Data Center A for Sol Terr Phys Boulder, CO

Kamide Y, Craven JD, Frank LA, Ahn B-H, Akasofu S-I (1986) Modeling substorm current systems using the conductivity distribution inferred from DE auroral images. J Geophys Res 91: 11235

Kamide Y, Ishihara Y, Killeen TL, Craven JD, Frank LA, Heelis RA (1989) Combining electric field and auroral observations from DE 1 and 2 with ground magnetometer records to estimate ionospheric electromagnetic quantities. J Geophys Res 94: 6723

Kan JR (1982) Towards a unified theory of discrete auroras. Space Sci Rev 31: 71

Kan JR (1984) Electrodynamics of magnetosphere–ionosphere coupling. In: Roederer JG (ed) Achievements of the IMS. ESA Sci Tech Pub Branch, Noordwijk, Netherlands, pp. 257

Kan JR (1987) Generation of field-aligned currents in magnetosphere–ionosphere coupling in a MHD plasma. Planet Space Sci 35: 903

Kan JR, Kamide Y (1985) Electrodynamics of the westward traveling surge. J Geophys Res 90: 7615

Kan JR, Sun W (1985) Simulation of the westward traveling surge and Pi 2 pulsations during substorms. J Geophys Res 90: 10911

Kan JR, Williams RL, Akasofu S-I (1984) A mechanism for the westward traveling surge during substorms. J Geophys Res 89: 2211

Kan JR, Zhu L, Akasofu S-I (1988) A theory of substorms: onset and subsidence. J Geophys Res 93: 5624

Kaufman RL (1984) What auroral electron and ion beams tell us about magnetosphere–ionosphere coupling. Space Sci Rev 37: 313

Kaufman RL, Kintner PM (1984) Upgoing ion beams. 2. Fluid analysis and magnetosphere–ionosphere coupling. J Geophys Res 89: 2195

Kawasaki K, Rostoker G (1979) Perturbation magnetic fields and current systems associated with eastward drifting auroral structures. J Geophys Res 84: 1464

Kaye SM, Kivelson MG (1979) Time-dependent convection electric fields and plasma injection. J Geophys Res 84: 4183

Kaye SM, Kivelson MG (1981) The influence of geomagnetic activity on the radial variation of the magnetospheric electric field between L = 4 and 10. J Geophys Res 86: 863

Kennel CF (1969) Consequences of a magnetospheric plasma. Rev Geophys 7: 379

Kern JW (1966) Analysis of polar magnetic storms. J Geomagn Geoelec 18: 125

Kindel JM, Kennell JF (1971) Topside current instabilities. J Geophys Res 76: 3055

Kirchhoff VW, Carpenter LA (1976) The day-to-day variability in ionospheric electric fields and currents. J Geophys Res 81: 2737

Kirkwood S, Opgenoorth HJ, Murphree JS (1988) Ionospheric conductivities, electric fields and currents associated with auroral substorms measured by the Eiscat radar. Planet Space Sci 36: 1359

Kisabeth JL (1979) On calculating magnetic and vector potential fields due to large-scale magnetospheric current systems and induced currents in an infinitely conducting earth. In: Olson WP (ed) Quantitative Modeling of Magnetospheric Processes. Am Geophys Union, Washington DC, pp. 473

Kisabeth JL, Rostoker G (1973) Current flow in auroral loops and surges inferred from ground-based magnetic observations. J Geophys Res 78: 5573

Kivelson MG (1976) Magnetospheric electric fields and their variation with geomagnetic activity. Rev Geophys Space Phys 14: 189

Klumpar DM, Burrows JR, Wilson MD (1976) Simultaneous observations of field-aligned currents and particle fluxes in the postmidnight sector. Geophys Res Lett 3: 395

Klumpar DM, Peterson WK, Shelley EG (1984) Direct evidence for two-stage (bimodal) acceleration of ionospheric ions. J Geophys Res 89: 10799

Knight S (1973) Parallel electric fields. Planet Space Sci 21: 741

Kremser G, Wilhelm K, Riedler W, Brønstad K, Trefall H, Ullaland SL, Legrand JP, Kangas J, Tanskanen P (1973) On the morphology of auroral-zone X-ray events. II. Events during the early morning hours. J Atmos Terr Phys 35: 713

Kunkel T, Baumjohann W, Untiedt J, Greenwald RA (1986) Electric fields and currents at the Harang discontinuity: A case study. J Geophys 59: 73

Küppers F, Untiedt J, Baumjohann W, Lange K, Jones AG (1979) A two-dimensional magnetometer array for ground-based observations of auroral zone electric currents during the International Magnetospheric Study (IMS). J Geophys 46: 429

Lanzerotti LJ, Regan RD, Sugiura M, Williams DJ (1976) Magnetometer networks during the International Magnetospheric Study. EOS Trans Am Geophys Union 57: 442

Lee L-C (1986) Magnetic flux transfer at the earth's magnetopause. In: Kamide Y, Slavin JA (eds) Solar Wind–Magnetosphere Coupling. Terra Reidel Dordrecht, pp. 297

Lemaire J, Scherer M (1974) Ionosphere-plasmasheet field-aligned currents and parallel electric fields. Planet Space Sci 22: 1485

Lennartsson W (1976) On the magnetic mirroring as the basic cause of paralle electric fields. J Geophys Res 81: 5583

Lester M, Hughes WJ, Singer HJ (1984) Polarization patterns of Pi 2 pulsations and the substorm current wedge. J Geophys Res 89: 5489

Levitin AE, Afonina RG, Belov BA, Feldstein YI (1982) Geomagnetic variation and field-aligned currents at northern high latitudes, and their relations to the solar wind parameters. Philos Trans R Soc Lond Ser A 304: 253

Lin CS, Rowland HL (1985) Anomalous resistivity and AE-D observations of auroral electron acceleration. J Geophys Res 90: 4221

Lin CS, Sugiura M, Burch JL, Barfield JN, Nielsen E (1984) DE 1 observations of type 1 counter-streaming electrons and field-aligned currents. J Geophys Res 89: 8907

Lockwood M (1984) Thermospheric control of the auroral source of O^+ ions for the magnetosphere. J Geophys Res 89: 301

Lockwood M, Waite JH Jr, Moore TE, Johnson JFE, Chappell CR (1985) A new source of suprathermal O^+ ions near the dayside polar cap boundary. J Geophys Res 90: 4099

Lu G, Reiff GP, Burch JL, Winningham JD (1991) On the auroral current–voltage relationship. J Geophys Res 96: 3532

Lui ATY, Anger CD (1973) A uniform belt of diffuse auroral emission seen by the ISIS 2 scanning auroral photometer. Planet Space Sci 21: 799

Lui ATY, Anger CD, Akasofu S-I (1975) The equatorward boundary of the diffuse aurora and auroral substorms as seen by the Isis 2 auroral scanning photometer. J Geophys Res 80: 3603

Lundin R (1976) Rocket observations of electron spectral and angular characteristics in an "inverted V" event. Planet Space Sci 24: 499

Lyatsky WB, Maltsev YP, Leontyev SV (1980) Three-dimensional current system in different phases of a substorm. Planet Space Sci 22: 1231

Lyon J, Brecht SH, Fedder JA, Palmadesso P (1980) The effects on the earth's magnetotail from shocks in the solar wind. Geophys Res Lett 7: 721

Lyons LR (1980) Generation of large-scale regions of auroral currents, electric potentials, and precipitation by the divergence of the convection electric field. J Geophys Res 85: 17

Lyons LR (1981) Discrete aurora as the direct result of an inferred high-altitude generating potential distribution. J Geophys Res 86: 1

Lyons LR (1983) Causes of particle precipitation along auroral field lines. In: Hultquist B, Hagfors T (eds) High-latitude Space Plasma Physics. Plenum Press, New York, pp. 493

Lyons LR, Walterschied RL (1985) Generation of auroral omega bands by shear instability of the neutral winds. J Geophys Res 90: 12321

Lysak RL (1985) Auroral electrodynamics with current and voltage generators. J Geophys Res 90: 4178

Lysak RL (1986) Coupling of the dynamic ionosphere to auroral flux tubes. J Geophys Res 91: 7047

Lysak RL, Dum CT (1983) Dynamics of magnetosphere–ionosphere coupling including turbulent transport. J Geophys Res 88: 365

Maekawa K, Maeda H (1978) Electric fields in the ionosphere produced by polar field-aligned currents. Nature 273: 649

Maezawa K (1976) Magnetospheric convection induced by the positive and negative Z components of the interplanetary magnetic field: Quantitative analysis using polar cap magnetic records. J Geophys Res 81: 2289

Maezawa K, Murayama T (1986) Solar wind velocity effects on the auroral zone magnetic disturbances. In: Kamide Y, Slavin JA (eds) Solar Wind–Magnetosphere Coupling. Terra Sci Reidel, Tokyo, pp. 59

Maier EJ, Kayser SE, Burrows JR, Klumpar DM (1980) The suprathermal electron contributions to high-latitude Birkeland currents. J Geophys Res 85: 2003

Maltsev YP (1974) The effect of ionospheric conductivity on the convection system in the magnetosphere. Geomagn Aeron 14: 128

Maral G, Brønstad K, Trefall H, Kremser G, Specht H, Tanskanen P, Riedler W, Legrand JP (1973) On the morphology of auroral-zone X-ray events. III. Large-scale observations in the midnight-to-morning sector. J Atmos Terr Phys 35: 735

Marklund GI (1984) Auroral arc classification scheme based on the observed arc-associated electric field pattern. Planet Space Sci 32: 193

Marklund G, Sandahl I, Opgenoorth H (1982) A study of the dynamics of a discrete auroral arc. Planet Space Sci 30: 179

Marshall JA, Burch JL, Kan JR, Reiff PH, Slavin JA (1991) Sources of field-aligned currents in the auroral plasma. Geophys Res Lett 18: 45

Matsushita S (1967) Electric conductivity. In: Matsushita S, Campbell WH (eds) Physics of Geomagnetic Phenomena. Academic Press, New York, pp. 381

Matsushita S, Mozer FS (1973) Origin of currents and electric fields in the dynamo region. Space Res 13: 397

Maynard NC (1974) Electric field measurements across the Harang discontinuity. J Geophys Res 79: 4620

Maynard NC (1978) On large poleward-directed electric fields at sub-auroral latitudes. Geophys Res Lett 5: 617

Maynard NC, Chen AJ (1975) Isolated cold plasma regions: observations and their relation to possible production mechanisms. J Geophys Res 80: 1009

Maynard NC, Grebowsky JM (1977) The plasmapause revisited. J Geophys Res 82: 1591

Mazaudier C (1985) Electric currents above Saint-Santin. 3. A preliminary study of disturbances: June 6, 1978; March 22, 1979; March 23, 1979. J Geophys Res 90: 1355

Mazaudier C, Blanc M, Nielsen E, Zi M-Y (1984) Latitudinal profile of the magnetospheric convection electric field at ionospheric altitude from a chain of magnetic and radar data. J Geophys Res 89: 375

McDiarmid IB, Burrows JR, Budzinski EE (1975) Average characteristics of magnetospheric electrons (150 eV to 200 keV) at 1400 km. J Geophys Res 80: 73

McPherron RL (1970) Growth phase of magnetospheric substorms. J. Geophys Res 75: 5592

McPherron RL (1974) Critical problems in establishing the morphology of substorms in space. In: McCormac BM (ed) Magnetospheric Physics. Reidel, Dordrecht, pp. 335

McPherron RL (1979) Magnetospheric substorms. Rev Geophys Space Phys 17: 657

McPherron RL (1991) Physical processes producing magnetospheric substorms and magnetic storms. In: Jacobs JA (ed) Geomagnetism, vol 4. Academic Press, New York, pp. 593–739

McPherron RL, Manka RH (1985) Dynamics of the 1054 UT March 22, 1979, substorm event: CDAW 6. J Geophys Res 90: 1175

Meng, C-I, Holzworth RH, Akasofu S-I (1977) Auroral circle: Delineating the poleward boundary of the quiet auroral belt. J Geophys Res 82: 164

Meng C-I, Snyder AL, Kroehl HW (1978) Observations of auroral westward traveling surges and electron precipitations. J Geophys Res 83: 575

Menietti JD, Burch JL (1985) Electron conic signatures observed in the nightside auroral zone and over the polar cap. J Geophys Res 90: 5345

Mersmann U, Baumjohann W, Küppers F, Lange K (1979) Analysis of an eastward electrojet by means of upward continuation of ground-based magnetometer data. J Geophys Res 45: 281

Mishin VM (1990) The magnetogram inversion technique and some applications. Space Sci Rev. 53: 83

Mishin VM, Bazarzhapov AD, Shpynev GB (1979) Electric fields and currents in the earth's magneto-sphere. In: Akasofu S-I (ed) Dynamics of the Magnetosphere. Reidel, Hingham, MA, pp. 249

Mitchell HG Jr, Palmadesso PL (1983) A dynamic model for the auroral field line plasma in the presence of field-aligned current. J Geophys Res 88: 2131

Miura A (1984) Anomalous transport by magnetohydrodynamic Kelvin-Helmholtz instabilities in the solar wind–magnetosphere interaction. J Geophys Res 89: 801

Miura A, Sato T (1980) Numerical simulation of global formation of auroral arcs. J Geophys Res 85: 73

Miura A, Sato T (1981) Global simulation of auroral arcs. In: Akasofu S-I, Kan JR (eds) Physics of Auroral Arc Formation. Am Geophys Union, Washington DC, pp. 321

Mizera PF, Fennell JF (1977) Signatures of electric field from high and low altitude particle distributions. Geophys Res Lett 4: 311

Mizera PF, Croley DR Jr, Fennell JF (1976) Electron pitch-angle distributions in an inverted V structure. Geophys Res Lett 3: 149

Mizera PF, Gorney DJ, Fennell JF (1982) Experimental verification of an S shaped potential structure. J Geophys Res 87: 1535

Moore TE, Chappell CR, Lockwood M, Waite JH Jr (1985) Suprathermal ion signatures of auroral acceleration processes. J Geophys Res 90: 1611

Mozer FS (1981) ISEE-1 observations of electrostatic shocks on auroral zone field lines between 2.5 and 7 earth radii. Geophys Res Lett 8: 893

Mozer FS (1984) Electric field evidence on the viscous interaction at the magnetopause, Geophys Res Lett 11: 135

Mozer FS, Lucht P (1974) The average auroral zone electric field. J Geophys Res 79: 1001

Mozer FS, Carlson CW, Hudson MK, Torbert RB, Parady B, Yatteau J, Kelley MC (1977) Obser-vations of paired electrostatic shocks in the polar magnetosphere. Phys Rev Lett 38: 292

Mozer FS, Cattell CA, Hudson MK Lysak RL, Temerin M, Torbert RB (1980) Satellite measure-ments and theories of low altitude auroral particle acceleration. Space Sci Rev 27: 155

Murayama T (1982) Coupling function between solar wind parameters and geomagnetic indices. Rev Geophys Space Phys 20: 623

Nagata T, Kokubun S (1962) An additional geomagnetic daily variation (Sq^p field) in the polar region on geomagnetically quiet days. Rep Ionos Space Res Jpn 16: 256

Nakamura R, Oguti T (1987) Drifts of auroral structures and magnetospheric electric fields. J Geophys Res 92: 11241

Nielsen E, Greenwald RA (1979) Electron flow and visual aurora at the Harang discontinuity. J Geophys Res 84: 4189

Nisbet JS, Miller MJ, Carpenter LA (1978) Currents and electric fields in the ionosphere due to field-aligned auroral currents. J Geophys Res 83: 2647

Nishida A (1966) Formation of plasmapause, or magnetospheric plasma knee by the combined action of magnetosphere convection and plasma escape from the tail. J Geophys Res 71: 5669

Nishida A (1968a) Geomagnetic DP 2 fluctuations and associated magnetospheric phenomena. J Geophys Res 73: 1795

Nishida A (1968b) Coherence of geomagnetic DP 2 fluctuations with interplanetary magnetic variations. J Geophys Res 73: 5549

Nishida A (1978) Geomagnetic diagnosis of the magnetosphere. Springer, Berlin Heidelberg New York

Nishida A (1979) Possible origin of transient dawn-to-dusk electric field in the nightside magneto-sphere. J Geophys Res 84: 3409

Nishida A, Kamide Y (1983) Magnetospheric processes preceding the onset of an isolated substorm—A case study of the March 31, 1978, substorm. J Geophys Res 88: 7005

Nopper R W Jr, Carovillano RL (1978) Polar-equatorial coupling during magnetically active periods. Geophys Res Lett 5: 699

Obayashi T, Nishida A (1968) Large-scale electric field in the magnetosphere. Space Sci Rev 8: 3

Oguti T (1976) Recurrent auroral patterns. J Geophys Res 81: 1782

Oguti T, Hayashi K (1984) Multiple correlations between auroral and magnetic pulsations. 2. Determination of electric currents and electric fields around a pulsating auroral patch. J Geophys Res 89: 7467

Oguti T, Meek JH, Hayashi K (1984) Multiple correlations between auroral and magnetic pulsations. J Geophys Res 89: 2295

Oguti T, Nakamura R, Yamamoto T (1987) Oscillations in drifts of auroral patches. J Geomagn Geoelec 39: 609

Ohtani S, Kokubun S, Elphic RC, Russell CT (1988) Field-aligned current signatures in the near-tail region. 1. ISEE observations in the plasma sheet boundary layer. J Geophys Res 93: 9709

Okuda H, Ashour-Abdalla M (1983) Acceleration of hydrogen ions and conic formation along auroral field lines. J Geophys Res 88: 899

Opgenoorth HJ, Pellinen RJ, Baumjohann W, Nielsen E, Marklund G, Eliasson L (1983a) Three-dimensional current flow and particle precipitation in a westward travelling surge (observed during the Barium-Geos rocket experiment). J Geophys Res 88: 3138.

Opgenoorth HJ, Oksman J, Kaila KU, Nielsen E, Baumjohann W (1983b) On the characteristics of eastward drifting omega bands in the morning sector of the auroral oval. J Geophys Res 88: 9171

Parks GK, McCarthy M, Fitzenreiter RJ, Etcheto J, Anderson KA, Eastman TE, Frank LA, Gurnett DA, Huang C, Lin RP, Lui ATY, Ogilvie KW, Pedersen A, Reme H, Williams DJ (1984) Particle and field characteristics of the high-latitude plasma sheet boundary layer. J Geophys Res 89: 8885

Paschmann G, Johnson RG, Sharp RD, Shelley EG (1972) Angular distributions of auroral electrons in the energy range 0.8–16 keV. J Geophys Res 77: 6111

Paschmann G, Sonnerup BUO, Papamastorakis I, Sckopke N, Haerendel G, Bame SJ, Asbridge JR, Gosling JT, Russell CT, Elphic RC (1979) Plasma acceleration at the Earth's magnetopause: Evidence for reconnection. Nature 282: 243

Pashin AB, Glassmeier K-H, Baumjohann W, Raspopov OM, Yahnin AG, Opgenoorth HJ, Pellinen RJ (1982) Pi 2 magnetic pulsations, auroral break-up, and the substorm current wedge: A case study. J Geophys Res 51: 223

Pellat R, Laval G (1972) Magnetospheric substorm phenomena. In: Dyer ER (ed) Critical Problems of Magnetospheric Physics. National Academy of Sciences, Washington DC, pp. 237

Pellinen RJ, Baumjohann W, Heikkila WJ, Sergeev VA, Yahnin AG, Marklund G, Melnikov AO (1982) Event study of pre-sub-storm phases and their relation to energy coupling between solar wind and magnetosphere. Planet Space Sci 30: 371

Perreault PD, Akasofu S-I (1978) A study of geomagnetic storms. Geophys J R Astr Soc 54: 547

Potemra TA (1987) Birkeland currents: Recent contributions from satellite magnetic field measurements. Phys Scripta T18: 152

Potemra TA, Zanetti LJ, Bythrow PF, Lui ATY, Iijima T (1984) B_y-dependent convection patterns during northward interplanetary magnetic field. J Geophys Res 89: 9753

Pudovkin MI (1974) Electric fields and currents in the ionosphere. Space Sci Rev 16: 727

Rees MH (1963) Auroral ionization and excitation by incident energetic electrons. Planet Space Sci 11: 1209

Rees MH (1982) Scientific results of the U.S. IMS ground-based program. Rev Geophys Space Phys 20: 654

Rees MH (1983) Auroral excitation and energy dissipation. In: Carovillano RL, Forbes (eds) Solar-terrestrial Physics. Reidel, Hingham, MA, pp 753

Reiff PH, Burch JL (1985) IMF B_y-dependent plasma flow and Birkeland currents in the dayside magnetosphere. 2. A global model for northward and southward IMF. J Geophys Res 90: 1595

Reiff PH, Spiro RW, Hill TW (1981) Dependence of polar cap potential drop on interplanetary parameters. J Geophys Res 86: 7639

Rich F, Kamide Y (1983) Convection electric fields and ionospheric currents derived from model field-aligned currents at high latitudes. J Geophys Res 88: 271

Rich FJ, Maynard NC (1989) Consequences of using simple analytical functions for the high-latitude convection electric field. J Geophys Res 94: 3687

Richmond AD (1976) Electric field in the ionosphere and plasmasphere on quiet days. J Geophys Res 81: 1447

Richmond AD Kamide Y (1988) Mapping electrodynamic features of the high-latitude ionosphere from localized observations: Technique. J Geophys Res 93: 5741

Richmond AD, Matsushita S, Tarpley JD (1976) On the production mechanism of electric currents and fields in the ionosphere. J Geophys Res 81: 547

Rishbeth H, Garriott OK (1969) Introduction to Ionospheric Physics. Academic Press, New York

Robinson RM (1984) Kp dependence of auroral zone field-aligned current intensity. J Geophys Res 89: 1743

Robinson RM, Bering EA, Vondrak RR, Anderson HR, Cloutier PA (1981) Simultaneous rocket and radar measurements of currents in an auroral arc. J Geophys Res 86: 7703

Robinson RM, Rich F, Vondrak RR (1985a) Chatanika radar and S3-2 measurements of auroral zone electrodynamics in the midnight sector. J Geophys Res 90: 8487

Robinson RM, Vondrak RR, Potemra TA (1985b) Auroral zone conductivities within the field-aligned current sheets. J Geophys Res 90: 9688

Robinson RM, Vondrak RR, Miller K, Dabbs T, Hardy D (1987) On calculating ionospheric conductances from the flux and energy of precipitating electrons. J Geophys Res 92: 2565

Rostoker G (1983) Triggering of expansive phase intensifications of magnetospheric substorms by northward turnings of the interplanetary magnetic field. J Geophys Res 88: 6981

Rostaker G, Boström R (1976) A mechanism for driving the gross Birkeland current configuration of the auroral oval. J Geophys Res 81: 235

Rostoker G, Hron MP (1975) The eastward electrojet in the dawn sector. Planet Space Sci 23: 1377

Rostoker G, Olson JV (1978) Pi2 pulsations as indicators of substrom onsets and intensifications. J Geomagn Geoelec 30: 135

Rostoker G, Samson JC (1981) Polarization characteristics of Pi2 pulsations and implications for their source mechanisms: Location of source regions with respect to auroral electrojets. Planet Space Sci 29: 225

Rostoker G, Samson JC (1984) Can substorm expansive phase effects and low frequency Pc magnetic pulsations be attributed to the same source mechanism? Geophys Res Lett 11: 271

Rostoker G, Akasofu S-I, Foster J, Greenwald RA, Kamide Y, Kawasaki K, Liu ATY, McPherron RL, Russell CT (1980) Magnetospheric substorms: Definition and signatures. J Geophys Res 85: 1663

Rostoker G, Mareshal M, Samson JC (1982) Response of dayside net downward field-aligned current to changes in the interplanetary magnetic field and to substrom perturbations. J Geophys Res 87: 3489

Rostoker G, Akasofu S-I, Baumjohann W, Kamide Y, McPherron RL (1987a) The roles of direct input of energy from the solar wind and unloading of stored magnetotail energy in driving magnetospheric substorms. Space Sci Rev 46: 93

Rostoker G, Vallance Jones A, Gattinger RL, Anger CD, Murphree JS (1987b) The development of the substorm expansive phase: The eye of the substorm. Geophys Res Lett 14: 399

Rothwell PL, Silevitch MB, Block LP (1984) A model for the propagation of the westward traveling surge. J Geophys Res 89: 8941

Rothwell PL, Silevitch MB, Block LP (1986) Pi2 pulsations and the westward traveling surge. J Geophys Res 91: 6921

Rothwell PL, Silevitch MB, Block LP, Tanskanen P (1988) A model of the westward traveling surge and the generation of Pi2 pulsations. J Geophys Res 93: 8613

Royrvik O, Davis TN (1977) Pulsating aurora: Local and global morphology. J Geophys Res 82: 4720

Russell CT (1987) The magnetosphere. In: Akasofu SI, Kamide Y (eds) The Solar Wind and the Earth. Reidel, Dordrecht, p 71

Russell CT, Elphic RC (1979) ISEE observations of flux transfer events at the dayside magnetopause. Geophys Res Lett 6: 33

Saito T (1978) Long-period irregular magnetic pulsation, Pi3. Space Sci Rev 21: 427

Saito T, Yumoto K, Koyama Y (1976) Magnetic pulsations Pi2 as a sensitive indicator of magnetospheric substorm. Planet Space Sci 24: 1025

Samson JC, Rostoker G (1983) Polarization characteristics of Pi2 pulsations and implications for their source mechanism: Influence of the westward traveling surge. Planet Space Sci 31: 435

Sato T (1978) A theory of quiet auroral arcs. J Geophys Res 83: 1042

Sato T (1982) Auroral physics. In: Nishida A (ed) Magnetospheric Plasma Physics. Center for Acad Publ/Reidel, Tokyo, pp 197

Sato T (1985) Principles of magnetohydrodynamic simulation in space plasmas. In: Matsumoto H, Sato T (eds) Computer Simulation of Space Plasmas. Kluwer, Hingham, MA, p 133

Sato T, Iijima T (1979) Primary sources of large-scale Birkeland currents. Space Sci Rev 24: 347

Sato T, Matsumoto H, Nagai K (1982) Particle acceleration in time-developing magnetic reconnection process. J Geophys Res 87: 6089

Sato T, Hayashi T, Walker RJ, Ashour-Abdalla M (1983) Neutral sheet current interruption and field-aligned current generation by three-dimensional driven reconnection. Geophys Res Lett 10: 221

Scholer M (1970) On the motion of artifical plasma clouds in the magnetosphere. Planet Space Sci 18: 977

Scholer M, Gloeckler G, Hovestadt D, Klecker B, Ipavich FM (1984) Characteristics of plasmoidlike structures in the distant magnetotail. J Geophys Res 89: 8872

Schulz M (1991) The magnetosphere. In: Jacobs JA (ed) Geomagnetism, vol 4. Academic Press, New York, pp 87–293

Scourfield MWJ, Keys JG, Nielsen E, Goertz CK, Collin H (1983) Evidence for the E × B drift of pulsating auroras. J Geophys Res 88: 7983

Segatz M (1985) Determination of two-dimensional ionospheric conductances and current system in the Harang discontinuity region by using combined observations of ground magnetic and ionospheric electric disturbances fields—A new method. Diploma Thesis, Institute of Geophysics, Univ of Münster, Munster, Germany

Senior C, Blanc M (1984) On the control of magnetospheric convection by the spatial distribution of ionospheric conductivities. J Geophys Res 89: 261

Senior C, Robinson RM, Potemra TA (1982) Relationship between field-aligned currents, diffuse auroral precipitation and the westward electrojet in the early morning sector. J Geophys Res 87: 10469

Senior C, Sharber JR, de la Beaujardiere O, Heelis RA, Evans DS, Winningham JD, Sugiura M, Hoegy WR (1987) E and F region study of the evening sector auroral oval: A Chatanika/Dynamics Explorer 2/NOAA 6 comparison. J Geophys Res 92: 2477

Shawhan ST, Fälthammer C-G, Block LP (1978) On the nature of large auroral zone electric fields at 1-R_E altitude. J Geophys Res 83: 1049

Shelley EG, Sharp RD, Johnson RG (1976) He^{++} and H^+ flux measurements in the dayside cusp: Estimates of convection electric field. J Geophys Res 81: 2363

Singh N, Schunk RW (1985) A possible mechanism for the observed streaming of O^+ and H^+ ions at nearly equal speeds in the distant magnetotail. J Geophys Res 90: 6361

Siscoe GL (1982a) Polar cap size and potential: A predicted relationship. Geophys Res Lett 9: 672

Siscoe GL (1982b) Energy coupling between regions 1 and 2 Birkeland current systems. J. Geophys Res 87: 5124

Siscoe GL (1988) The magnetospheric boundary. In: Chang T, Crew GB, Jasperse JR (eds) Physics of Space Plasmas. Scientific Publ, Cambridge, MA, p 3

Slavin JA, Baker AD, Craven JD, Elphic RC, Fairfield DH, Frank LA, Galvin AB, Hughes WJ, Manka RH, Mitchell DG, Richardson IG, Sanderson TR, Sibeck DJ, Smith EJ, Zwickl RD (1989) CDAW 8 observations of plasmoid signatures in the geomagnetic tail: An assessment. J Geophys Res 94: 15153

Smiddy M, Kelley MC, Burke W, Rich F, Sagalyn R, Shuman S, Hoys R, Lai S (1977) Intense poleward-directed electric fields near the ionospheric projection of the plasmapause. Geophys Res Lett 4: 543

Sonnerup BUO (1971) Adiabatic particle orbits in a magnetic null sheet. J Geophys Res 76: 8211

Sonnerup BUO (1980) Theory of the low-latitude boundary layer. J Geophys Res 85: 2017

Southwood DJ (1977) The role of hot plasma in magnetospheric convection. J Geophys Res 82: 5512

Southwood DJ, Wolf RA (1978) An assessment of the role of precipitation in magnetospheric convection. J Geophys Res 83: 5227

Spiro RW, Wolf RA (1983) Electrodynamics of convection in the inner magnetosphere. In: Potemra TA (ed) Magnetospheric Currents. Am Geophys Union, Washington DC, pp 247

Spiro RW, Harel M, Wolf RA, Reiff PH (1981) Quantitative simulation of a magnetospheric substorm. 3. Plasmaspheric electric fields and evolution of the plasmapause. J Geophys Res 86: 2261

Spiro RW, Reiff PH, Maher LJ (1982) Precipitating electron energy flux and auroral zone conductivities—An empirical model. J. Geophys Res. 87: 8215

Stasiewicz K (1984) On the origin of the auroral inverted-V electron spectra. Planet Space Sci 32: 379

Stenbaek-Nielsen HC, Hallinan TJ, Wescott EM, Foeppl H (1984) Acceleration of barium ions near 8000 km above an aurora. J Geophys Res 89: 10788

Stern DP (1975) The motion of a proton in the equatorial magnetosphere. J Geophys Res 80: 595

Stern DP (1977) Large-scale electric fields in the Earth's magnetosphere, Rev Geophys Space Phys 15: 156

Stern DP (1981) One-dimensional models of quasi-neutral parallel electric fields. J Geophys Res. 86: 5837

Stern DP (1983) The origins of Birkeland currents. Rev Geophys Space Phys 21: 125

Stiles GS, Foster JC, Doupnik JR (1980) Prolonged radar observations of an auroral arc. J Geophys Res 85: 1223

Sugiura M (1972) Equatorial current sheet in the magnetosphere. J Geophys Res 77: 6093

Sugiura M (1975) Identifications of the polar cap boundary and the auroral belt in the high-altitude magnetosphere: A model for field-aligned currents. J Geophys Res. 80: 2057

Sugiura M, Maynard NC, Farthing WH, Heppner JP, Ledley BG, Cahill LJ (1982) Initial results on the correlation between the electric and magnetic fields observed from the DE 2 satellite in the field-aligned current regions. Geophys Res Lett 9: 985

Sugiura M, Iyemori T, Hoffman RA, Maynard NC, Burch JL, Winningham JD (1984) Relationships between field-aligned currents, electric fields, and particle precipitation as observed by Dynamics Explorer 2. In: Potemra TA (ed) Magnetospheric Currents. AGU, Washington, pp. 96–103

Sun W, Kan JR (1985) A transient-response theory of Pi2 pulsations. J. Geophys. Res. 90: 4395

Sun W, Lee L-C, Kamide Y, Akasofu S-I (1985) An improvement of the Kamide–Richmond–Matsushita scheme for the estimation of the three-dimensional current system from ground magnetometer data. J Geophys Res 90: 6469

Swift DW (1967) Possible consequences of the asymmetric development of the ring current belt. Planet Space Sci. 15: 835

Swift DW (1968) Further possible consequences of the asymmetric development of the ring current belt—Effect of variations in ionospheric conductivity. Planet Space Sci 16: 329

Swift DW (1971) Possible mechanisms for formation of the ring current belt. J Geophys Res. 76: 2276

Tamao T (1986) Direct contribution of oblique field-aligned currents to ground magnetic fields. J Geophys Res 91: 183

Tanaka T (1986) Low-latitude ionospheric disturbances: Results for March 22, 1979, and their general characteristics. Geophys Res Lett 13: 1399

Tarpley JD (1970) The ionospheric wind dynamo. 1. Lunar tide. Planet Space Sci 18: 1075

Taylor HE, Perkins FW (1971) Auroral phenomena driven by the magnetospheric plasma. J Geophys Res 76: 272

Temerin M, Lysak RL (1984) Electromagnetic ion cyclotron mode (ELF) waves generated by auroral electron precipitation. J Geophys Res 89: 2849

Temerin MA, Boehm MH, Mozer FS (1981) Paired electrostatic shocks. Geophys Res Lett 8: 799

Testud J, Amayenc P, Blanc M (1975) Middle and low latitude effects of auroral disturbances from incoherent scatter. J Atmos Terr Phys 37: 989

Tighe WG, Rostoker G (1981) Characteristics of westward traveling surges during magnetospheric substorms. J Geophys Res 50: 51

Tiwari MS, Rostoker G (1984) Field-aligned currents and auroral acceleration by non-linear MHD waves, Planet Space Sci. 32: 1497

Troshichev OA (1982) Polar magnetic disturbances and field-aligned currents. Space Sci. Rev 32: 275

Troshichev OA, Kuznetsov BM, Pudovkin MI (1974) The current systems of the magnetic substorm growth and explosive phase. Planet Space Sci 22: 1403

Tsunoda RT, Presnell RI, Potemra TA (1969a) The spatial relationship between the evening radar aurora and field-aligned currents. J Geophys Res 81: 3891

Tsunoda RT, Presnell RI, Kamide Y, Akasofu S-I (1976b) Relationship of radar auroral, visual aurora and auroral electrojets in the evening sector. J Geophys Res 81: 6005

Tsurutani BT, Thorne RM (1982) Diffusion processes in the magnetopause boundary layer. Geophys Res Lett 9: 1247

Untiedt J, Segatz M, Kuerschner M, Glaßmeier K-H (1990) Direct determination of the local ionospheric Hall conductance distribution from two-dimensional electric and magnetic field data. J Geophys Res (submitted)

Vasyliunas VM (1970a) Mathematical models of magnetospheric convection and its coupling to the ionosphere In: McCormac BM (ed) Particles and Fields in the Magnetosphere. Reidel, Hingham, MA

Vasyliunas VM (1970b) Models of ionospheric currents driven by quasi-static magnetospheric convection and their relation to equivalent current systems. Paper presented at the Upper Atmospheric Currents and Electric Fields Symp, Natl Center for Atmos Res and Environ Sci Serv Admin, August, 1970b, Boulder, CO

Vasyliunas VM (1972) The interrelationship of magnetospheric processes. In: McCormac BM (ed) Earth's Magnetospheric Processes. Reidel, Hingham, MA, pp 29

Vasyliunas VM (1975) Theoretical models of magnetic field line merging. 1. Rev Geophys Space Phys 13: 303

Vickrey JF, Vondrak RR, Mathews SJ (1981) The diurnal and latitudinal variation of auroral zone ionospheric conductivity. J Geophys Res 86: 65

Vickrey JF, Livingston RC, Walker NB, Potemra TA, Heelis RA, Kelley MC, Rich FJ (1986) On the current–voltage relationship of the magnetospheric generator at intermediate spatial scales. Geophys Res Lett 13: 495

Volland H (1973) A semiempirical model of large-scale magnetospheric electric fields. J Geophys Res 78: 171

Vondrak RR (1983) Incoherent scatter radar measurements of electric field and plasma in the auroral ionosphere. In: Hultqvist B, Hagfors T (eds) High Latitude Space Plasma Physics. Plenum Press, New York, pp 73

Vondrak RR, Rich FJ (1982) Simultaneous Chatanika radar and S3-2 satellite mesurements of ionospheric electrodynamics in the diffuse aurora. J Geophys Res 87: 6173

Wagner JS Kan JR (1985) On the field-aligned scale length of the V-shaped auroral potential structure. Planet Space Sci. 33: 89

Wagner JS, Lin CS, Tajima T (1985) Simulation study of type 2 counterstreaming electrons along auroral field lines. J Geophys Res 90: 4249

Waite JH Jr, Nagai T, Johnson JFE, Chappell CR, Burch JL, Killeen TL, Hays PB, Carignan GR, Peterson WK, Shelley EG (1985) Escape of suprathermal O^+ ions in the polar cap. J Geophys Res 90: 1619

Wallis DD, Budzinski EE (1981) Empirical models of height integrated conductivities. J Geophys Res 86: 125

Wand RH (1981) A model representation of the ionospheric electric field over Millstone Hill ($= 56°$). J Geophys Res 86: 5801

Watanabe K and Sato T (1988) Self-excitation of auroral arcs in a three-dimensionally coupled magnetosphere–ionosphere system. Geophys. Res Lett 15: 717

Watanabe K, Ashour-Abdalla M, Sato T (1986) A numerical model of magnetosphere–ionosphere coupling: Preliminary results. J Geophys Res 91: 6973

Wedde T, Doupnik JR, Banks PM (1977) Chatanika observations of the latitudinal structure of electric fields and particle precipitation on November 21, 1975. J Geophys Res 82: 2743

Weimer DR, Gurnett DA, Goertz CK, Menietti JD, Burch JL (1985) Auroral zone electric fields from DE 1 and 2 at magnetic conjunctions. J Geophys Res 90: 7479

Weimer DR, Gurnett DA, Goertz CK, Menietti JD, Burch JL, Sugiura M (1987) The current–voltage relationship in auroral current sheets. J Geophys Res 92: 187

Weimer DR, Maynard NC, Burke WJ, Liebrecht C (1990) Polar cap potentials and the auroral electrojet indices. Planet Space Sci 38: 1207

Wescott EM, Peek HM, Stenbaek-Nielsen HC, Murcray WB, Jensen RJ, Davis TN (1972) Two successful field line tracing experiments. J Geophys Res. 77: 2982

Wescott EM, Stenbaek-Nielsen HC, Hallinan TJ, Davis TN, Peek HM (1976) The Skylab barium plasma injection experiments, 2, Evidence for a double layer. J Geophys Res. 81: 4495

Wilhjelm J, Friis-Christensen E, Potemra TA (1978) The relation between ionospheric and field-aligned current in the dayside cusp. J Geophys Res 83: 5586

Winckler JR, Steffen JE, Malcolm PP, Erickson KN, Abe Y, Swanson RL (1984) Ion resonances and ELF wave production by an electronic beam injected into the ionosphere: Echo 6. J Geophys Res. 89: 7565

Winninham JD, Yasuhara F, Akasofu SI, Heikkila WJ (1975) The latitudinal morphology of 10-eV to 10-keV electron fluxes during magnetically quiet and disturbed times in the 2100-0300 MLT sector. J Geophys Res 80. 3148

Wolf RA (1970) Effects of ionospheric conductivity on convective flow of plasma in the magnetosphere. J Geophys Res. 75: 4677

Wolf RA (1974) Calculations of magnetospheric electric fields. In: McCormac BM (ed) Magneto-spheric Physics. Reidel Dordrecht, pp 167

Wolf RA (1975) Ionosphere–magnetosphere coupling. Space Sci Rev 17: 537

Wolf RA (1983) The quasi-static (slow-flow) region of the magnetosphere. In: Carovillano RL, Forbes JM (eds) Solar-terrestrial Physics. Reidel, Hingham, MA, pp 303

Wolf RA, Kamide Y (1983) Inferring electric fields and currents from ground magnetometer data—a test with theoretically derived input. J Geophys Res 88: 8129

Wolf RA, Spiro RW (1984) Ionosphere–magnetosphere coupling and convection. In: Roederer JG (ed) Proc Conf Achievements of the IMS. ESA Sci Tech Pub Branch, Noordiwijk, Netherlands, pp 417

Wolf RA, Mantjoukis GA, Spiro RW (1986) Theoretical comments on the nature of the plasmapause. Adv. Space Res 6(3): 177

Wu C-C, Walker RJ, Dawson JM (1981) A three-dimensional MHD model of the earth's magneto-sphere. Geophys Res Lett 8: 528

Wygant JR, Torbert RB, Mozer FS (1983) Comparison of S3-3 polar cap potential drops with the interplanetary magnetic field and models of magnetopause reconnection. J Geophys Res. 88: 5727

Yamamoto T, Kan JR (1985) The field-aligned scale length of one-dimensional double layers. J Geophys Res 90: 1553

Yasuhara F, Kamide Y, Akasofu S-I (1975) Field-aligned and ionospheric currents. Planet Space Sci 23: 1355

Yasuhara F, Greenwald R, Akasofu S-I (1983) On the rotation of the polar cap potential pattern and associated polar phenomena. J Geophys Res 88: 5773

Yasuhara F, Kamide Y, Vickrey JF (1985) On the efficiency of the Cowling mechanism in the auroral elecrojet. Geophys Res Lett 12: 389

Yau AW, Shelley EG, Peterson WK, Lenchyshyn L (1985) Energetic auroral and polar ion outflow at DE 1 altitudes: Magnitude, composition, magnetic activity dependence, and long-term varia-tions. J Geophys Res 90: 8417

Yeh H-C, Hill TW (1981) Mechanism of parallel electric fields inferred from observations. J Geophys Res 86: 6706

Zanetti LJ, Baumjohann W, Potemra TA (1983) Ionospheric and Birkeland current distributions inferred from the Magsat magnetometer data. J Geophys Res 88: 4875

Zanetti LJ, Potemra TA, Iijima T, Baumjohann W, Bythrow PF (1984) Ionospheric and Birkeland current distributions for northward interplanetary magnetic field: Inferred polar convection. J Geophys Res 89: 7453

Zi M, Nielsen E (1982) Spatial variation of electric fields in the high-latitude ionosphere. J Geophys Res 87: 5202

Zi M-Y, Shen C-S (1986) A time-dependent analytical model with day-night ionospheric conductivity gradient for magnetospheric convection. Plan Space Sci 34: 353

Ziesolleck S, Baumjohann W, Brüning K, Carlson CW, Bush RI (1983) Comparison of height-integrated current densities derived from ground-based magnetometer and rocket-borne obser-vations during the Porcupine F3 and F4 flights. J Geophys Res 88: 8063

Zmuda AJ, Armstrong JC (1974a) The diurnal variation of the region with vector magnetic field changes associated with field-aligned currents. J Geophys Res 79: 2501

Zmuda AJ, Armstrong JC (1974b) The diurnal flow pattern of field-aligned currents. J Geophys Res 79: 4611

Zmuda AJ, Martin JH, Heuring FT (1966) Transverse magnetic disturbances at 1100 kilometers in the auroral region. J Geophys Res 71: 5033

Subject Index